WITHDRAWN

Anonymous Gift

REASONABLE USE

REASONABLE USE

The People,
the Environment,
and the State,
New England 1790–1930

John T. Cumbler

UNIVERSITY PRESS

2001

OXFORD
UNIVERSITY PRESS

Oxford   New York
Athens   Auckland   Bangkok   Bogotá   Buenos Aires   Calcutta
Cape Town   Chennai   Dar es Salaam   Delhi   Florence   Hong Kong   Istanbul
Karachi   Kuala Lumpur   Madrid   Melbourne   Mexico City   Mumbai
Nairobi   Paris   São Paulo   Shanghai   Singapore   Taipei   Tokyo   Toronto   Warsaw

and associated companies in
Berlin   Ibadan

Copyright © 2001 by Oxford University Press, Inc.

Published by Oxford University Press, Inc.
198 Madison Avenue, New York, New York 10016

Oxford is a registered trademark of Oxford University Press.

All rights reserved. No part of this publication may be reproduced,
stored in a retrieval system, or transmitted, in any form or by any means,
electronic, mechanical, photocopying, recording, or otherwise,
without the prior permission of Oxford University Press.

Library of Congress Cataloging-in-Publication Data
Cumbler, John T.
Reasonable use : the people, the environment, and the state, New England, 1790–1930/
John T. Cumbler.
p. cm.
Includes bibliographical references and index.
ISBN 0-19-513813-9
1. Nature—Effect of human beings on—Connecticut River Valley.   2. Human ecology—Connecticut River Valley.   3. Industrialization—Connecticut River Valley.   4. Environmental degradation—Connecticut River Valley.   5. Environmental policy—Connecticut River Valley.   6. Connecticut River Valley—Environmental conditions   I. Title.
GF504.C65 C85 2000
333.77'14'0974—dc21   00-050159

1 3 5 7 9 8 6 4 2
Printed in the United State of America
on acid-free paper

To Sam Bass Warner,
who taught me the past,
and Kazia and Ethan,
who are the future

Acknowledgments

Many people have provided me time, energy, patience, living space, and financial support for this work, without which it would never have been completed or improved. The weaknesses that remain are no fault of theirs but rather are due to my pigheadedness and impatience. Judith Cumbler patiently put up with the long hours and anxiety that went into my completing this work. Lew Erenberg, Harold Platt, Jim O'Brien, Richard Judd, and Edward Countryman read all or most of this manuscript and gave me more than just helpful advice. Lew and Harold were particularly generous with their time. Malcolm Fleschner also read over the complete manuscript and did his best to help me improve my style. Mark Blum read over the complete manuscript and offered kind encouragement. All along the way, people have helped me with bits and pieces of this work. Some I know, and others were anonymous reviewers. Sam Hays, Joel Tarr, Mary Blewett, Nancy Theriot, Ian and Astrid Fletcher, Chad Montrie, Bill Dakan, Steve Miller, Sally Benbasset, and Ethan Cumbler all contributed to this work. Mary Hawkesworth was especially helpful. Joyce Berry and Jessica Ryan helped get the manuscript in shape for public viewing. Other people lent me their couches and beds—Rusty Scudder, Libby and Bruce Bartolini, Milton Cantor, Charles Stevenson, and Gail and Walter Willett. Numerous librarians, archivists, and editors contributed their time and support to this project. A generous fellowship from the National Endowment for the Humanities gave me the time needed to complete this book. And finally, Sam Bass Warner not only read the manuscript more than once and encouraged me and provided important support and insight, he also continues to be my mentor, my teacher, and my friend. For that reason, he shares the dedication with my children.

Contents

Introduction:
The Environment, the People, and the State:
The Connecticut River Valley, 1790–1930   3

1   The Land, the River, and the People:
    The Connecticut Valley, 1790–1830   11

2   From Milling to Manufacturing: From Villages to Mill Towns   33

3   Cities and Industry, Sewage and Waste   49

4   Pre-1860 Responses to Change: Views of the Public Good   63

5   Fish, the People, and Theodore Lyman:
    The Moderate Approach   79

6   Health, State Medicine, and Henry Ingersoll Bowditch:
    The Radical Approach   103

7   Cooperation, Conflict, and Reaction   119

8   Industrial Waste, Germs, and Pollution: The Battle
    over Pollution   131

9   "Most Beautiful Sewer"   147

10 Farmers, Fishers, and Sportsmen   161

11 New England, the Nation, and Us   181

    Notes   193

    Index   257

# REASONABLE USE

Introduction

*The Environment, the People, and the State:
The Connecticut River Valley, 1790–1930*

Early twentieth-century conservation in the United States has been identified in the public mind with the West and the protection of wilderness, parks, and national forests.[1] Some scholars have explored conservation through the writings of naturalists and antimodernists like Henry David Thoreau. What we have only recently come to appreciate is that there was a whole generation of reformers very much concerned about the environment who were neither antimodernists nor wilderness protectors.[2] They were modernists who rejected not the modern world, but the way the modern world was being fashioned. They did not retreat or long to retreat into the wilderness but lived in cities and towns. And they struggled to make the environment of the most settled parts of the nation more amenable to human habitation.

It was in New England where these reformers first began to make their claims for the rights of citizens to clean air, clean water, and clean soil.[3] The Massachusetts board of health argued, less than five years after the Civil War, for aggressive state action on the claim that "all citizens have an inherent right to the enjoyment of pure and uncontaminated air, and water, and soil, that this right should be regarded as belonging to the whole community, and that no one should be allowed to trespass upon it by his carelessness or his avarice."[4] And the New Hampshire board, in its first report, stated that "every person has a legitimate right to nature's gifts—pure water, air, and soil—a right belonging to every individual, and every community upon which no one should be allowed to trespass through carelessness, ignorance, or other cause."[5]

New England's first environmental crisis was brought on by its people's fecundity and by their material practices in the late eighteenth century.[6] Out of that crisis emerged a changed New England with concentrated manufacturing centers and increasingly market-oriented agriculture. Although not all New Englanders enthusiastically supported this change, all were affected by it. Within three generations, New Englanders saw their region transformed. That transformation created a new set of troubles. The emergence of those new problems, and the solutions nineteenth-century Yankees offered, is the story of this book.

The story begins in 1790, when much of the land of the Connecticut River Valley, especially away from the lowland settled areas, had abundant wildlife, clean clear water, and forest-covered hills and mountains. In the lowlands, the land was rich in fertile soil. By the third decade of the nineteenth century, local merchants and traders, as well as Boston investors, were damming the falls along New England's waterways for transportation access and power to run machinery.[7] In 1848, Boston capital helped build the dam at Holyoke, Massachusetts, that harnessed the power of the Connecticut River to run the textile machines in the mills, encouraging the building of both mills and the tenements that housed those who worked in them. These investments transformed the sleepy farming village of South Hadley, Massachusetts—where farmers gathered in the spring to fish for shad and salmon—into the industrial city of Holyoke, where workers gathered at mill gates. The New England world at midcentury was increasingly a world of machines and of cities like Holyoke, where nature was harnessed and controlled in ever newer and more radical fashions. By the 1860s, rivers such as the Connecticut, once sources of fish, as well as arteries of transportation, took on new roles in a world focused not around growing crops, harvesting lumber, and shipping out surpluses, but around powering machinery, washing and processing raw materials, and disposing of wastes. In such a world, once clear, clean, flowing water became increasingly stagnant and polluted. Hills and mountains were being rapidly stripped of their forest cover, while wildlife—finned, feathered, and four legged—was vastly depleted. The scenic river valleys filled with smoky, dreary, industrial cities that dumped pollution and wastes into the region's rivers and ponds and covered the landscape with tenements, mills, canals, and railroads.

Just as the ecological crisis of the late eighteenth century was the result of human actions, decisions, and processes of production and reproduction, so its solving too would be a product of human agency. Industrialization was the result of merchants and traders investing capital and mobilizing resources.[8] It was also the result of farmers and Irish immigrants and farmers' sons and daughters hiring out their labor.

For some New Englanders, the transformation of their region meant prosperity and wealth. For others, it meant disease and deprivation. For all New Englanders, it meant change, and many believed they needed to control, moderate, or at least understand that change. For Yale University president Timothy Dwight, who traveled the region at the turn of the century, the change represented the coming of civilization and the taming of nature. For the historian and failed farmer and trader Sylvester Judd, it meant the passing of the more innocent time. For naturalist and philosopher Henry David Thoreau, New England's change represented a loss of purity, independence, and virtue; for his companion naturalist George Perkins Marsh, it foreshadowed an environmental crisis.

Those who invested their capital and directed the construction of dams and mills understood they were transforming their physical world, but they did not share the pessimistic beliefs of Thoreau, Marsh, or even Judd. Theodore Lyman III's family was part of that cluster of Boston investors who with their allies in the local communities provided the capital to build the dams and mills and construct industrial towns such as Chicopee and Holyoke.[9] Like others of his set, Lyman believed manufacturing resolved the problem of decline in New England. As he noted in 1882 when he campaigned for Congress, "As long as Massachusetts was overlaid by 10 feet of gravel, she would have to manufacture or starve."[10] Investors like Lyman believed they were acting for the greater good: Their mills and dams provided employment and led to prosperity for the region. But their investments also increased their personal wealth and had to be defended politically and intellectually.

These capitalists created an environment in which rivers and streams were polluted by wastes dumped from the mills and tenements, fish were excluded from spawning grounds, and mill towns filled with smoke and foul odors. A generation removed from the original investors, Theodore Lyman III celebrated the prosperity of industrialization, but he realized that his family and their associates had also helped create a world of "prosaic faces, . . . long hard streets, ugly buildings, and sickly smells."[11]

Others did not feel the new world was as wonderful as did Theodore Lyman and his associates. Small marginal farmers who were dependent upon the "commons" (resources collectively owned and used particularly spring fish runs), as well as commercial fishers, found that dams and pollution reduced fish runs. Farmers whose fields were spoiled by polluted, flooded streams and whose cattle were made sick by fouled waters agitated against the degraded environment. Workers, middle-class urban residents, and emergent professionals armed with the latest in scientific knowledge were also concerned about the quality of this new urban world. Those who wanted to mitigate the environmental consequence of the urban industrial world confronted the power of those who profited from its expansion and would potentially bear the costs of reform. Out of that confrontation emerged a prolonged political struggle over power, authority, and the rights to the commons as either a sink or a resource. For the most part, manufacturers held to the idea that industrialization itself was a good that involved certain necessary social costs. They projected themselves as both the defenders of the common good and as agents of prosperity and progress. They fought any challenge to the rights of manufactures. At times, they were joined by industrial communities that like the manufacturers enjoyed the privilege of dumping their waste sewage into the nearest running water and were concerned to maintain industrial employment.

Those who attempted to ameliorate environmental degradation thus had to confront the power of both the manufacturers and urban communities. That confrontation drew a variety of reformers and supporters of reform. Some, such as scientist and public servant Theodore Lyman III or technocrat Hiram Mills, believed that it was possible through science and technology to moderate and direct New England's environmental alteration while retaining its benefits and its privileges. Lyman's and Mills's moderate position had many supporters, for it gave rise to far less opposition from the manufacturers and promised both economic and environmental progress. Although it prefigured the Progressives, who would adopt a similar strategy a generation later, it had its shortcomings. Science and technology could do only so much without eventually confronting the privilege of the polluters, who were unwilling to compromise if doing so would significantly hurt profits or generate too great a cost. If Lyman's position was the moderate one, doctor and reformer Henry Ingersoll Bowditch pushed for a radical approach. Bowditch saw environmental degradation as something to be confronted and ameliorated for the benefit of the poorest and weakest.

These New Englanders did not use the language of ecology and environmentalism, but they did have an awareness of their environment, and they attempted to understand and deal with the changes occurring around them. Their response involved a different notion of the public good. Their concerns about the environment as a public good and the struggle to achieve that good play a central role in this story. The new industrialized and urbanized landscape also engendered changing perceptions of science and technology, another focus of this book. Finally, New Englanders grappled with new instruments, particularly the power of the state, to ameliorate and control the byproducts of urbanization and industrialization. That is a third theme in this book.

The struggle of these nineteenth-century New Englanders has much to tell us today, for they confronted the beginnings of our own world. By understanding where they succeeded, where they failed, and how they understood this new world, we can better understand our own confrontation with modern industrial society.

This study will look at New England's environmental change, particularly in the Connecticut River Valley, and at how the people of the region understood that change and responded to it between 1790 and 1930. The Connecticut River, in a metaphoric sense, will be the narrator of this story. During the eighteenth and early nineteenth centuries, the river was bordered by farm and forest. But this was a world already in the throes of change: Forests were cleared, land was more intensively farmed, and rural mill towns sprang up. To carry more goods in and out, the river was molded and shaped by canals and locks. The Civil War found the river dammed and increasingly bordered by industrial cities; by the second half of the nineteenth century, its condition was the focus of concern not only for the river's immediate neighbors, but also for distant reformers and legislators, who become part of the river's story. By the twentieth century, upland farms had reverted to forest, tourists and sports hunters and fishers wandered the wilder regions of the river valley, and cities and suburbs defined much of the river's less remote world. As this progression suggests, the history of the valley falls roughly into three periods. Section one of this book focuses on the first period, the process of industrialization. The second centers around the response to that industrializing process. The third deals with the rise of the mediating state and the successes and failures of that response.

If the river is the metaphoric narrator of this work, fish are the indicators of the narrator's health. They are the river's way of telling people

Connecticut River Valley. Holyoke and Connecticut Rivers, Holyoke, Massachusetts.

about the well-being of its ecosystem. But that telling gets translated though the individuals who lived by the river, studied it, cared for it, or cared about it. The translators' voices and understanding are also part of this story. They are the river's human voice.

All historical time periods are somewhat arbitrary, chosen by the historian. This book ends at 1930 because by then the patterns of New England's environmental change and response were established. The 1930s also launched a new era in the region as the federal government became a major player in its environment. In 1927, the new River and Harbor Act involved the Army Corps of Engineers and the Federal Power Commission in planning flood control and dam building in New England and elsewhere. In 1933, the New England Regional Planning Commission began systematically charting the region's river basin.[12]

The setting for this book is New England, generally, and particularly the Connecticut River Valley. Because the Connecticut River runs through four of the six New England states and its watershed touches such a large part of the region's area, the valley is an appropriate place to find examples of the larger story I want to tell. The Connecticut was New England's largest and most important waterway. The river was a primary transportation artery and a source of power for a significant number of the region's industries. Its prolific runs of fish also provided an important resource for the locals and for export. The Connecticut River Valley also encompasses within it much of what is typical of New England: tidal commercial cities (Middletown and Hartford), inland industrial cities (Chicopee and Holyoke), fertile lower valleys, and forests and farms of the upriver hill country.

For many New Englanders, the Connecticut was a symbol of the region, both in its beauty and its utility. The loss of fish in the river (and in other New England rivers), the increased pollution, and the environmental degradation inspired farmers, fishers, and urban residents to pressure their various state legislatures to do something. That pressure also mobilized reformers throughout the region, including those who lived in the metropolis, in the struggle for fish and game restoration and for public health reform to clean up the physical environment.

Although this study goes to the Connecticut River Valley for its details, much of the legislative and legal history it recounts centers on the state of Massachusetts. Massachusetts gets the lion's share of attention because many of the related legislative and ideological battles were first fought there. Massachusetts then became a model for the other New England states. But this remains a study of New England, for what was happening in Massachusetts was also happening, although usually a little later, in the other New England states—states linked by the Connecticut River Valley's ecology. Many of the nineteenth- and early twentieth-century battles over pollution and the environment concerned rivers and water systems, yet this is not only a story about New England's waters, it is also a story about the entire New England environment. Decisions about fish and water pollution were ultimately decisions about how to manage the environment itself.

The story of the struggle to maintain clean waters is also about politics in the broadest sense. Conflict—political, economic, social, and intellectual—permeates this book. The people in this story defended their interests, whether they were industrialists, farmers, fishers, health reformers, tenement dwellers, or casual summer recreationists. Environmental history must report this struggle if it is to have any meaning for today.

Much progress has been made in the field of environmental history. Recently, it has come to focus more attention on the actual physical landscape and the constraints of nature on human action. Environmental history has also moved beyond the wilderness to embrace the urban world, the role of technology, and the evolution of science. In the last quarter of the twentieth century, race and class played a more prominent role in books and articles in the field.[13] Yet while environmental historians have broadened their agenda, the central politics of conflict is often neglected in their work. As most environmental historians understand, the natural world does impact and constrain human agency. But human agency itself changes and forms the natural world. And the form that this human agency takes emerges out of struggle. It is a struggle that most often involves access to resources, and for whom and how these resources are to be used. That is what this book is about. Although specific to one place within one region, it is also very much a history of America. To tell that history it is necessary to cross intellectual boundaries. Although primarily a social-environmental history, it borrows heavily from the fields of industrial, political, intellectual, legal, and urban history.

The environmental history of New England is important not just as regional history. It is also important because New England was a pioneer in the urban industrial transformation that would influence so much of the development of this nation. As New Englanders struggled with the environmental change occurring about them, they developed an understanding and a discourse of conservation that influenced the understanding and language of later conservationists.

New Englanders, confronting waters depleted of fish and an environment polluted by wastes, realized that their problems were not just local or state problems, but regional ones. The common meetings in the 1860s and 1870s of the New England Commissioners of River Fisheries to work out joint policy and to coordinate activity evolved out of necessity, and even as this policy faltered, it became a model for other regional commissions confronting interstate environmental problems.

It is easy to lionize famous conservationists such as John Muir, who called the nation's attention to the need to protect the wilderness. But in the second half of the nineteenth century, a host of New Englanders—working people, industrialists, political leaders, and public health officers—were struggling with greater problems: how to protect dwindling resources and declining quality of environmental life within a highly developed region and how to protect resources that were already compromised. For New Englanders, the problems concerned protecting the quality of life for Americans who lived not in cabins in wilderness valleys, but in homes and tenements in urban centers. Increasingly, that was where most Americans dwelt.

1

The Land, the River, and the People
The Connecticut Valley, 1790–1830

The Connecticut Valley

On Wednesday morning September 21, 1795, only a year after he was appointed president of Yale College, forty-four-year-old Timothy Dwight began the first of his thirteen excursions through New England and upstate New York. On six of his thirteen trips, he traveled through the Connecticut Valley, a valley he was familiar with since childhood and was linked to by both family and sentiment.

The Connecticut River Valley was changing, as Dwight made his several trips through it. It was transformed under the impact of human activity. Increasingly, mill dams and factory villages were being built along the river and its tributaries. Technology, science, and the market were restructuring the way people were interacting with their environment. The land became less wild. That "civilizing" of nature, as Dwight called it, began first on the alluvial soils of the lower and central valley in the eighteenth century and then spread north and up into the hill country in the early years of the nineteenth century. By the end of the fifth decade of the nineteenth century, this new world had pretty much taken shape, and valley residents began to take stock of the changes that had occurred. Dwight began this process of accounting at the beginning stages of that transformation. And it was in the Connecticut River Valley that the changes made the biggest impact on him.

At the center of the Connecticut Valley runs New England's largest waterway. The Connecticut River flows south some four hundred miles from a series of small lakes in the swampy district of northern New Hampshire on the Canadian border. It eventually spills into Long Island Sound at Saybrook, Connecticut.[1] To the west and east of the river are mountain ranges, the Housatonic and Green Mountains to the west and the White Mountains to the east. In northern New Hampshire and Vermont, the river travels through a narrow and rough mountain valley.[2] As the river moves south into central Vermont and New Hampshire, the valley widens, particularly on the river's western shore, and is intersected with tributary rivers and valleys. High mountains frame this upper part of the valley and give it impressive vistas. The river flows from southern Vermont and New Hampshire down into its rich central valley through Massachusetts to Enfield Falls at the Connecticut border. Here the valley opens up to an average width of forty miles, and the river meanders through the remnants of two wide ancient Ice Age lakes to Middletown, Connecticut. The basin here is mostly of alluvial formation. Several major hills rise up from this lowland valley, the most impressive being the Mount Holyoke Range and the Mount Tom Range, just north of Holyoke.[3]

Along most of its path, the river flows through a broad valley resting on a bed of Triassic rock. As the Wisconsin Glacier, which covered much of the valley during the Pleistocene period, retreated, it left behind clay and silt outwash.[4] Over time, the river in its broad valleys eroded out a series of terraces. The terraces rise from the river in successive magnificent steps. The lower steps consist of rich alluvial meadows or intervals; while the higher ones are remnants of ancient floodplains later overflowed by glacial rivers.[5] A number of major tributaries also flow into the Connecticut.[6] At Middletown, the river turns southeast through a gorge and flows over a belt of crystalline rock and through a narrow channel with steep sides to Saybrook and Lyme on the Long Island Sound.

If, unlike Dwight, who traveled north from Middletown, one were to travel west from Boston to the Connecticut and then move north and south up and down the river valley, one would metaphorically capture the valley's history. Boston was the metropolis that held the valley within its reach. Although valley goods traveled south by river and then by ocean to distant ports, it was to Boston that area residents looked for political guidance and authority. Boston was the source of much of the capital for the region's development. The fate of the valley, even those parts outside the jurisdiction of the Massachusetts legislature, was heavily influenced by affairs in Boston. But even with Boston's heavy hand and the importance of rail links between the valley and the commonwealth's capital, the valley was also linked north and south by the river itself. Lumber from the mountains of Vermont and New Hampshire flowed downriver to build dams, mills, and tenements in the towns to the south, while food from the region's farms fed the growing commercial cities and mill towns. The Connecticut River held the people of the region's largest watershed in its watery grasp and linked them to a common history.

Although when Dwight made his journey, the forests of the lower valleys were already heavily cut, the steeper mountains and hills as well as much of the upper valley were still "buried in . . . forest" "with innumerable tall trees."[7] And even though Dwight believed that with "the rapid progress of cultivation, the hills, plains, and valleys of [the upper valley would] be stripped of the[ir] forest [cover]," in 1797 the region was still "majestically and even gloomily overshadowed" with forest.[8] The northern valley had only a few scattered farms on recently cleared land. When Dwight made his trip up the Connecticut, turkey, partridge, quail, passenger pigeon, and waterfowl, as well as fish, were plentiful.[9]

The farms Dwight passed on his ride north rested on brown podzolic soil. Brown podzolics are fairly acidic and easily leached. They required constant fertilization and liming. The soil above the valley floor on the bluffs or plateaus, the upper terraces, suffered particularly from acidity. Yet compared to much of the rest of New England, the Connecticut Valley's lowland alluvial soil was particularly fertile.

Timothy Dwight saw the transformation that encompassed the valley in positive terms. Typical of someone of his background and training, Dwight was an empiricist and naturalist by inclination. It became his habit to take copious notes on his travels, commenting on everything from geography, topography, history, and natural history to the flora and fauna of the areas he passed through. In his notes, Dwight wrote of the theme of civilization overcoming wilderness through the virtue of Yankee ingenuity and hard work. He believed in a controlled nature, in which "lands literally useless" were transformed into a "garden." He wanted his Yankee farm neighbors to practice modern scientific agriculture. Dwight also came to see that the publication of his notes would document the changes occurring in New England.[10]

Dwight saw two different rural New Englands in his travels. One included the developed, commercially oriented agricultural and trading communities of the lower and central valley, home to artisans, traders, and prosperous farmers holding their land in fee simple and growing wheat, oats, corn, and rye.[11] The other was the older and more traditional New England found on the hill farms and in the upper reaches of the valley. Although this one was not self-sufficient, it was subsistence based, dependent upon cooperation and sharing with neighbors, fishing, hunting, free-range pasturing, and long fallow periods.[12]

A typical hill farm of fifty or more acres would have over one-third of the arable land left fallow. Another third of the land would be devoted to woodlots to be harvested to heat the home, used for building, or sold off for other supplies.[13] Dwight believed that farmers who took up farms on the hills and upper valley were content to live the "half-working, half-lounging life" of marginal subsistence.[14] These farmers would clear a section of the forest for corn and flax and allow their animals to browse in the forests and on "bush pastures." Supplemented by hunting and fishing, this farming would keep families alive, but it provided few of the trappings of the cultivated life that Dwight found so admirable.[15] Despairing about the character of these northern hill farmers, Dwight complained that they had "a deficiency in the labor necessary to prepare the ground for seed, insufficient manuring, the want of a good rotation of crops, and slovenliness in clearing the ground."[16]

One of the traditional practices of upland farmers was to let their livestock run free in the woods. Pigs and sheep, particularly, but also horses and cattle, would be marked by the farmer and then left to fend for themselves in the woods and meadows. Ebenezer Kingsley told Sylvester Judd that in the 1780s, "cattle got their living in the woods."[17] These old ways of the poorer farmers led to conflict between them and their more prosperous neighbors. In 1806, the wealthier farmers placed a notice in the *Hampshire Gazette* informing those continuing to leave their animals free that the "people living at South Farms . . . were determined henceforward to put a stop to the unjust and unchristian practice of too many people . . . to turn their sheep and cattle upon our farms." The South Farms farmers "have united to prevent the destruction of their woodlots, pastures, and fences."[18]

Increasingly, as conflicts emerged between farmers whose animals ran free and those who fenced their land, the courts became inundated with cases of animals "breaking into an enclosed close" and damaging crops or pasturage. Settled prosperous towns began to enforce laws requiring animals to be fenced.[19]

Farms and Farm Families

Farmers used the land differently according to where and when they farmed, but whatever their circumstances, it was the land itself that they had to manage in order to survive. Despite Dwight's complaints that nature's fecundity encouraged indolence, times were not easy for these farmers. Crops failed. Hard winters or early frosts, heavy rains, flooding, droughts, or pests could wipe out a farmer's food supply. Those who survived these calamities managed partly through luck and partly through borrowing and trading with neighbors and family.[20]

These families ate the bounty of their farms.[21] For bedding, these families gathered pigeon feathers.[22] Passenger pigeons flew over New England in prodigious numbers. The huge flocks of pigeons, which darkened the sky for days, were not a yearly occurrence but would arrive irregularly, usually once every seven or eight years, in what were known as "pigeon years."[23] The plentitude of pigeons had its drawbacks; poor Levi Moody complained that he ate "pigeon every day for a month or two."[24]

Valley families at the end of the eighteenth and early nineteenth centuries supplemented the food they grew or raised with more than just pigeons. The resources of the commons, wild animals, fish, and wild nuts and fruits were vital parts of these families' diets.[25] Indeed, Dwight complained that "food provided by game from the forest and fish from the lakes and streams" sustained larger families than could be maintained by what they grew and raised themselves and encouraged indolence.[26] Ducks and geese also added variety to the New England diet, so much so that the ancient charters of New England (which were the statutes of the earlier colonial period and became the basis for the new state laws and continued as rights into the national period) ensured access to all great ponds in order to "fish and fowl."[27]

The Water's Resources

Of all the indigenous fauna, fish were the most important to eighteenth- and early-nineteenth-century New Englanders, and not only the fish in the ocean, but also those in the inland rivers, streams, and ponds. Fish provided residents with an important part of their diet and a medium of exchange.[28] At a hearing in 1865, it was estimated that before the dams, people of the river valleys "derived about one-third of their annual food from [the rivers]."[29] As a Massachusetts court noted in 1807, "Fish furnished food to the inhabitants," and their destruction was deemed a "great injury of the public, an evil example of all others in like cases offending against the peace and dignity of the commonwealth."[30] Indeed, access to the fish of New England was deemed so important that (as the courts stated in *Vinton v. Welsh*) "constant legislation upon this subject [the regulation and protection of fish] from the first settlement of the country . . . subject[s] it to the control of the legislature in the manner and to the extent it has been immemorially exercised."[31]

In the eighteenth century, salmon, shad, herring, bass, and eels came up the Connecticut by the hundreds of thousands. Elihu Warner reminisced to Sylvester Judd that salmon weighing as much as thirty to forty pounds were caught "in great numbers." So thick with fish were the river and its tributaries that "it seemed as though a person could almost walk across [the river] upon their backs."[32] For Dwight, concerned about improvement and progress, the fishers lacked "enterprise," but when the shad came upriver, the fishers themselves swarmed to the riverbanks for fishing and socializing. Sometime shortly after the second week in April, the shad would appear at Hadley Falls, and the run would last for the next ten to twenty days. "Immense numbers [of people] came after the fish—all with horses and bags. . . . They caught on both sides of the river at the foot of the falls. . . . [Levi Moody] says many went for a frolic as much as for fish. Many stayed for 2 or 3 days who might have got their fish in one day. There was much drinking and frolicking and some card playing. Taverns and houses were all full."[33] Fishing times were periods of socializing for farmers after a long period of winter isolation. The fishers "indulged in plays and trials of skill. Where there were so many men, and rum was plenty, there was of course much noise, bustle and confusion."[34] These men could afford to frolic because when the rivers were so full of fish their catching was not a difficult enterprise. In the eighteenth and early nineteenth centuries, fishers along the Connecticut pulled in thousands of fish during a single day.[35]

Although people socialized as they fished, much to the displeasure of Dwight, fishing was their purpose, for fish supplied food for their families, especially in the spring when there was "a deficiency of pork."[36] "The greater part were industrious farmers and after leaving the falls, they wound over the hills and plains with bags of shad, in every direction. They were plainly dressed according to their business."[37]

Farmers in the central part of the valley had been shipping surplus goods downriver since the seventeenth century, but the focus of much farm activity was on production of food and goods for consumption within the farm family or local community. Within that system of production and consumption, there was a continual circulation of goods, labor, and services among neighbors.[38] Transformation also occurred. Goods were processed. Lumber was cut. Flax was cured, dressed, bleached, and spun. Wool had to be carded, spun, woven, and fulled. Grain needed to be ground and sifted, beef and pork butchered, dressed, salted or dried, and leather tanned. Some of that work occurred on the farm. Some took place at the sawmill, gristmill, fulling mill, or tannery. Much of the processing was done by the local miller or the storekeeper, who collected a good bit of the surplus goods of the community through his or her central place in this system of exchange and circulation. Not all goods that changed hands went through the country store, but the surplus or extra of the region did tend to end up there.[39]

Even though farm families exchanged goods and labor among themselves, families still needed some goods that could not be locally produced.[40] Country stores took in a wide variety of things produced on the farm, some of which shopkeepers sold to others in the area, and the rest of which they tried to sell to larger merchants and traders in the mercantile centers of Boston and Hartford. For example, in the late eighteenth century, G. Lyman and Levi Shepard accepted and sold flax, suet, mutton, pigeon, wheat, boards, turnips, rye, honey, clover seed, butter, beef, tow cloth [homespun cloth made from flax], flour, beeswax, corn, oats, wood, cider, wool, salmon, hay, salt pork, and labor.[41] When a farmer needed hard cash for taxes or dowry items from Boston, he turned again to the country store owner, who would usually be willing to pay cash for goods more easily saleable in long-distance markets.[42]

Country Stores and Market Produce

Pressure on rural farmers to focus on market crops and on improvement agriculture came from larger demographic and economic forces that were building up in the New England countryside. These slowly pulled more and more farm families away from traditional practices and into more land-intensive, market-oriented farming. The trend spread up into the hills and north up the valley, but it was a slow process covering a generation or more in which valley farmers gradually abandoned practices to which they were accustomed.[43]

By the late 1780s, store owners were more anxious to receive goods that they could easily sell for cash or credit in the larger market. One valley store owner told his customers in 1789 that he preferred specific kinds of commodities. "Cash and a generous price given for all kinds of shipping furs and beeswax, and part cash for pork and English grain."[44] Although store owners may have preferred certain items in exchange for the goods they sold, or even cash, getting them from the surrounding farmers proved difficult.[45] In 1786, Breck, Shepard, and Clark's country store outside Springfield offered "a general assortment [of goods] which [they] proposed to sell on the most reasonable of terms for ready pay in cash, grain, pork, potash, salts, flax-seed, etc., etc." in order to keep their traditional customers.[46] Tappan and Fowle ran a notice to farmers in the central valley that they had a variety of manufactured goods, clothes, silks, ribbons, tea, sugar, chocolate, spices, and so on: "Country produce of most kinds will be received in payment for the above goods."[47] Prescott and Dexter, as new store owners anxious to bring in customers, offered goods in exchange for most country produce.[48]

Levi Shepard of Northampton announced to the local community that he was willing to sell a large assortment of European and India goods "at a very small profit for cash." Although Shepard wanted cash, he also accepted "in payment most kinds of country produce, especially flax which is well dressed for the duck manufacturer."[49]

Farmers may have preferred the old system of exchange and trade, but merchants increasingly demanded that goods be paid for either in cash or by short-term credit or contracts for specific country produce. Seth Wright announced that he had a variety of goods from New York, Boston, England, and the West Indies that he would sell at the lowest price for cash. Yet Wright, reflecting the older patterns, also "continue[d] to take many articles of country produce for goods as usual."[50] What may have been "usual" in 1790 was increasingly not so as the decade progressed.[51] By the end of 1791, Levi Shepard, who two years earlier accepted for "payment most kinds of country produce," now announced that "goods could be paid for by either cash or 3 months credit, or on contracts for flax."[52]

Although, by the 1790s, country stores in the central valley only took cash, short-term credit, or specific highly marketable goods, in Brattleboro, thirty miles north of Northampton and above Turners Falls, storekeepers still accepted most "country goods" through the early years of the nineteenth century.[53]

By the early nineteenth century, merchants north of Turners Falls also sold for cash or specified what "country produce" they would take. In 1803, Richard Bigelow had "a variety of summer goods" for sale that he was willing "to sell cheap for ready pay" or "for those who have not cash to spare, good butter and tow cloth will be received in payment."[54] By 1805, Bigelow took only "rye" at four shillings a bushel, but he offered that in "English and West Indian goods," not cash.[55] Cyrus Chapin in late 1808 announced that he would sell "cheap as can be purchased in this state," but only for "ready cash." Unfortunately, Cyrus had to compete for customers with Oliver Chapin and Son, who opened a new store "ready to sell a variety of goods both English and West Indian which they determined to sell at the lowest terms for cash, rye, corn, or short approved credit."[56] After several months of holding out for cash without takers, Cyrus was forced to offer his goods for "cash or in exchange for rye or flax-seed."[57]

By the end of the second decade of the nineteenth century, store owners north of Turners Falls were also moving to a cash system.[58] After receiving "a new supply of goods," Hays and Hubbard in 1818 offered them "below the Boston retail prices for cash or short credit."[59] S. Cutting announced that he would "sell as low as his neighbor even if below Boston prices and the public may be assured that no one will or can sell goods lower than he will for cash or short and approved credit."[60] Merchants and, increasingly into the nineteenth century, manufacturers encouraged farmers to bring specific goods for which they would be paid in cash.[61]

Although the Connecticut River Valley's soils, particularly in the lowlands, were fertile, they were also subject to depletion. The region's soils consisted of a layer of organic material made up of decayed plant and animal matter and live insects, worms, and microbes, and a lower thin layer of leached minerals. The soil's acidity could be neutralized by calcium, magnesium, or potash. Constantly cleared fields and continuous planting increased the soil's acidity, which required farmers to add gypsum or lime to the earth. Continuous planting also depleted the soil of nutrients and humus (organic material), forcing farmers to increase inputs to the soil.[62]

Merchants' pressure on farmers to produce goods that were more marketable was not the only influence on farm families. Maintaining even a relatively self-sufficient farm required the constant labor of many hands. Just cutting twelve to fifteen cords of wood to heat the home required hours of labor, let alone milking and caring for animals, making and repairing tools, planting and harvesting crops, slaughtering and preparing meat, making cloth, clothes, candles, and soap, putting up vegetables, baking bread and making meals, hauling water, and cleaning out the home and barn. Farm families coped with these chores by spreading out the labor among many within the family. The more children in the family, the more the labor could be spread out. But children grew up, and as sons and daughters reached adulthood, they also needed farms and dowries. Sons required land for their own farms or cash for apprenticeships and setting up in a trade. The land for farms could come from either dividing up the family farm or providing cash for children to move on to new areas. New England farmers pursued both strategies, and both forced a radical transformation of farming.

New England farms declined in size in the early nineteenth century.[63] Dividing up the family farm meant that the land on the new farms had to be utilized more efficiently. Fields no longer could rest fallow for several years but needed more intense labor to keep them productive with manure and careful crop rotation.[64] Woodlots had to be more carefully managed, or the farm would soon be stripped of all its fuel supply. More careful attention had to be paid to pastures and to the sowing and harvesting of meadow grasses.[65] Dowries for daughters, trades for sons, and the sending of sons to new farms farther up the Connecticut River Valley into Vermont and New Hampshire or west into New York and Ohio also required that the family farm itself become more focused on marketable agricultural goods.[66]

Long-distance markets had always influenced local farmers, particularly farmers along the lower Connecticut Valley, who had easy transportation out of the region for their goods by river. By the end of the eighteenth century, however, more and more farmers felt the pressure of those markets on their day-to-day decisions. The West Indies and East Coast mercantile cities wanted dried or salted beef and pork, grains, and nonperishable vegetables. In response, farmers planted more onions, rye, or wheat and increased the numbers of cattle and hogs on the farm.[67]

In the fall of 1789, Sylvester Judd Sr., a justice of the peace, storekeeper, and farmer in Hadley, gathered up his and his neighbors' surplus cattle and drove them off to the Boston market. In doing so, he reduced the number of animals he had to feed over the winter and picked up some extra cash. In 1818, his son, the historian Sylvester Judd Jr., sent his cattle to market in the spring after fattening them up in the barn over the winter. For the younger Judd, the sale of cattle meant not extra cash but rather the survival of the farm. The cattle he sent to market were exclusively his, raised from the beginning with an eye on the market price for beef.[68]

The shift in farming practices between the two generations of Judds reflects a general shift that occurred throughout the valley between the late eighteenth and early nineteenth centuries. That shift affected the way people interacted with their surrounding countryside. By the turn of the century, cattle that once had roamed free were carefully tended now in fenced pastures.

Farmers, in determining what to plant and how to manage the farm, calculated not only the needs of their families but also what the storekeepers would accept. Storekeepers helped the farmers in making these decisions by offering contracts for goods. Levi Shepard, who expanded his store to include a duck-manufacturing enterprise to produce linen cloth, began offering contracts at his store for flax. Farmers short of cash could contract with Shepard to produce flax. This would guarantee Shepard flax for his duck manufacturing and gave the farmer a source of credit at the store. "Shepard requests those gentlemen that are in the farming business who have made contracts for the delivery of flax, that they would dress it out as early as possible that he may be enabled to pursue his duck manufacture the next season without interruption as a few interruptions that is occasioned for want of stock to keep the workmen employed will totally discourage the prosecution of this useful manufacture in the country. In short the whole encouragement of our manufacturers in general depend upon the farming interest, and of consequence the prosperity and happiness of the community at this time is at their disposal."[69]

Levi Shepard was not the only merchant opening up a manufacturing enterprise on the basis of his mercantile operations. B. Prescott not only sold a "large and neat assortment of English and India goods" but also manufactured "wool and cotton cards" and opened a malt works.[70] North country store owners of the Bellows Falls Company at the turn of the century sold "English [and] . . . West India goods" and ran "wool carding machines" in exchange for a portion of the wool of surrounding farmers.[71]

Although the reformers lamented the indifference of hill-town farmers to the new farming methods, the language and imagery of the promoters of modern agricultural methods and the pressure of market agriculture began to penetrate the northern and more remote areas.

The *Vermont Republican and American Yeoman* in 1818 pleaded to its readers to improve their agricultural practices. The paper felt that "unless farmers are fully impressed with the necessity of attending to the principle that it is the business of agriculture to improve instead of impoverish land, it will be needless to point out any modes of improving land as they would be disregarded and neglected as they have heretofore been."[72] For the reformers, the shifts to market agriculture and scientific farming were linked. The *Connecticut Valley Farmer and Mechanic* noted that "if the farmer should sell the hay, corn, potatoes, beef, pork, broom corn, and tobacco which he raises, and take in pay an equivalent in such articles as the foregoing . . . how long would the soil of his farm be worth cultivating?"[73] According to the *Connecticut Valley Farmer and Mechanic*, farmers had to practice modern scientific farming in order to keep their farms productive and profitable.

But the reformers had a hard time convincing the traditional farmers, who did not view farming activity as "the business of agriculture" but rather as a means of sustaining a family. Vermont and New Hampshire farmers, and farmers in the Massachusetts hill towns up from the prosperous valley floor, resisted the new agriculture.[74] Working on more marginal hilly land, these farmers continued to practice extensive cultivation, opening up new fields rather than manuring older ones, and abandoning worn-out fields.[75] But by the second decade of the nineteenth century, these traditionalists were hard-pressed to maintain their way of life. In 1816, "the cold year, or famine year," frost covered the ground every month of the year, and it snowed in July in much of New England; the following year, although not as harsh as 1816, was also hard. Crops failed throughout the region, and hay was in short supply. Animals died for lack of food, and farm families that usually raised all their own food found themselves short. The cold years of 1816 and 1817 broke the back of many small, more subsistence-based farmers.[76]

Although distant markets made the biggest initial impact on commercially oriented agriculture in New England, other influences such as Levi Shepard's "prosecution of . . . useful manufacture in the country" and the growth of urban inland cities themselves soon came to dominate commercial agriculture.[77]

Flax, which had traditionally been grown for family use to produce tow cloth for summer clothes, increasingly was in demand by local manufacturers like Levi Shepard.[78] And local manufacturers were not the only ones who sought valley farmers' flax. In 1790, the *Hampshire Gazette* urged farmers to take notice of the opening of sailcloth factories on the coast and to plant flax. "The price it now bears will afford ample encouragement to those who are disposed to raise it."[79]

Although flax commanded the attention of the *Hampshire Gazette* in 1790, by the early nineteenth century, farmers oriented toward manufacturing markets shifted to hemp, which proved a better material for cloth.[80]

Early-nineteenth-century New England farmers also raised sheep for wool for local manufacturers. By the third decade of the nineteenth century, northern New England, particularly Vermont, had become a major center for wool production, with over twenty-three thousand sheep in New England raised mostly for local mills.[81] Unfortunately for northern valley farmers, wool prices came down in the late 1840s as cheaper western wool took over the market. Increasingly after midcentury, valley farmers abandoned sheep for dairy cattle.

Different farm crops and methods were not the only changes occurring in the region. When farmers opened up new fields or gave their sons unimproved sections for new farms, this land was usually forested. Forest cover needed to be cleared to open up either new fields or new farms. The forest itself provided the farmer with marketable commodities; lumber or potash, which was used to make lye for soap, glass, or gunpowder, was easily traded at the country store for goods or cash. Burning a few acres of forest would produce bushels of potash, for which local store owners would pay up to forty dollars for a hundred bushels.[82] Lumber could be sold for cord wood, split for staves for barrels, or sold to sawmills. Once cleared, fields could either be used for pasture or plowed for crops. Either way, the ecology of the area was significantly changed. Without forest cover to protect moss-banked rivers and streams, the moss died back, banks were more exposed, and water temperature rose. Rainfall on the hills was more likely to run off into streams and creeks rather than be absorbed into the ground. Deforestation had an impact on river flooding as early as 1797, when Timothy Dwight noted at Bellows Falls that "the river now is often fuller than it probably ever was before the country above was cleared of its forests, the snows in open ground melting much more suddenly and forming much greater freshets than in forested ground."[83] Theodore Dwight observed in 1830 that spring and autumn floods increasingly flooded the Connecticut River Valley floor.[84]

Milk, butter, and cheese destined for local urban communities and the distant markets of Boston and New York encouraged farmers to keep more and more dairy cows, and to keep them in milk over the winter.[85] By the middle decades of the nineteenth century, whole regions, particularly of the northern valley, began to specialize in dairy farming. Dairy farming and keeping cows milking twelve months a year (as opposed to the eighteenth-century custom of letting the cows go dry over the winter) concentrated more animals in smaller areas. It also meant more attention to winter feed. More milking cows meant more muck, and more attention and manure on hay fields. Agricultural reformers encouraged farmers to manure their fields, and with more winter feeding, farmers had more manure to spread. The mucking of fields became increasingly common in the nineteenth century.[86] The increased use of manure increased per-acre yields but also increased nitrates and salts in the soil.

Despite the pessimism of the agricultural reformers over the practices of upland farmers, between Timothy Dwight's first trip through the region and his last in 1815, the countryside of New England had changed dramatically. By his last trip, Dwight found that particularly in the lower alluvial lands, farmers, linked to wider markets by navigable waterways, were already using more scientific methods of farming, importing newer, more efficient agricultural implements, rotating crops, and planting ground cover. Their "fences were better made" and their animals less likely to be let free to forage.[87] Dwight saw the change of the lower valley farmers as a shift from "desert to a garden," where villages have risen up and "lands literally useless are made to yield sustenance and convenience to mankind."[88]

The valley of improved farms around Hadley, Northampton, Hatfield, and Deerfield presented Dwight with a pleasing vista of "a more vigorous, enterprising, [and] improved husbandry," with "superior" farms "in prosperous circumstances."[89] For Dwight, there was "something far more delightful [than palaces] in contemplating the diffusion of enterprise and industry over an immense forest."[90] Increasingly, farmers' relationship to the land was mediated by science, technology, and the market. But other agents besides farmers were also changing the landscape.

Grinding Grain and Sawing Lumber

On January 5, 1780, Matthew Patten, a New Hampshire farmer, lawyer, and justice of the peace, filled two bags with grain and rode through the snow to Samuel Moor's mill to have his grain ground into flour. Patten depended upon the mill, for without it, he and his family would have had to spend hours grinding their grain by hand. The mill also sawed lumber for Patten. Millers like Moor were scattered about the late-eighteenth- and early-nineteenth-century New England countryside. And New Englanders saw them as vital for their survival.[91]

The need to grind cereal and saw lumber to build houses, barns, and fences led to the construction of dams to power gristmills and sawmills. By the middle of the seventeenth century, tributaries of the Connecticut had been dammed by boulders, earth, logs, and wooden planks to create heads of water and races that would direct water to underthrow mill wheels. Although these mills were under constant threat from ice, floods, fire, and collapsing dams, their importance to the community led to their continual rebuilding.[92]

By the end of the fourth decade of the nineteenth century, the nature of New England mills had changed. Increasingly, mills were not where millers ground or sawed, but where operators tended machines. The new mills were still located on dam sites and kept the term "mill," but now they were turning driveshafts rather than millstones. The transformation from grinding and sawing to running machines moved through a series of uneven steps. But before these New Englanders built the large textile and paper mills and machine shops, they built gristmills and sawmills, and these early mills were seen as vital to the community.

New England legislatures, understanding the significance of these gristmills and sawmills, passed a series of bills known as the Mill Acts (1795 c74, 1797 c63, and the colonial acts that preceded these) to encourage the building of mills. When the Massachusetts legislature passed these acts, New England was a rural society of farmers, tradespeople, and artisans. Mills were important enough that in 1805, when Seth Spring's mill dam flooded Sylvanus Lowell's fields, Judge Sedgwick of the Supreme Judicial Court of Massachusetts noted that "sacred rights of private property are never to be invaded but for obvious and important purposes of public utility. Such are all things necessary to the upholding of mills."[93] Reflecting the same idea four decades later, Judge Samuel Hubbard of the supreme judicial court noted in 1846 that "originally, when the inhabitants were few, and their means of erecting expensive works small, the erection of a mill was a great public benefit, and those who were willing to incur the expense were considered as public benefactors."[94] Mills were considered, as Chief Justice Shaw stated in 1839, a "better use of the water power upon considerations of public policy and the general good."[95] The mills initially built under these acts served local farmers who brought their grain, wood, or wool to be processed.[96] Once processed, the worked goods would be brought back to the farm to be baked into bread, built into fences, homes, or barns, or sewn into clothes.[97]

Although mills and mill dams were important for the local community, they were also the source of conflict. One conflict between mill owners and local residents was over fish. Common law and court decisions defended people's rights to fish both for food and for revenue.[98] Fish were "articles of food and traffic to [the] citizens," as John Loring testified to the 1865 Massachusetts joint legislative commission looking into fish depletion.[99]

When the fish were running, local farmers and traders abandoned their fields and shops, posted "gone fishing" signs, and headed for the rivers to catch salmon, shad, and alewives.[100] Fish runs up the major rivers of New England sustained a significant fishing industry of their own.[101] Recognizing the importance of this natural resource, the colonial general court shortly after initial settlement passed a series of statutes designed to protect fish populations and access to them. Of particular concern for the early settlers were the anadromous fish, those that migrated up the rivers and streams. The colonists believed these fish were vulnerable to having their migrations cut off by obstructions, either fish weirs or dams.

Overfishing by a predator, either human or nonhuman, could dangerously reduce stocks. In the wild, fish are free, and following the logic of John Locke, no one can claim them until someone expends the labor to capture them.[102] During their migrations to their spawning grounds, anadromous fish move in thick concentrated schools over a short period of time.[103] Any one person or group of persons could take advantage of this fact. In doing so, they could reap huge short-term benefits while at the same time risking the continuity of the stock.[104]

New England settlers were anxious both to protect access to the fisheries for members of the community and to protect the resource from depletion. That balancing act was accomplished by a series of statutes passed while settlements still hugged the eastern seaboard. The Massachusetts Bay Colony passed laws protecting fishing and fowling rights to all citizens on any navigable waterway and on any "great pond" of over ten acres.[105] Fishing rights on waters above the navigable point rested with the riparian proprietor (or those capturing the right through prescription), "subject, however, like such rights acquired under legislative grant, and all other rights of fishing, either in the sea or in rivers, to regulation by legislature, so as to prevent the obstruction of the passage of fish."[106] But while the colonies extended fishing rights to all freeholders to any navigable waterway and great pond, they also limited fishing in order to protect the resource. By the eighteenth century, the colony passed laws preventing weirs "or other disturbance or incumbrance . . . on or across any river, to the stopping, obstructing . . . the natural or usual course and passage of fish in their seasons or spring of the year."[107] The colony did allow that "nothing herein contained shall be construed to extend to the pulling down or demolishing of any mill-dam already made, or that shall hereafter be lawfully and orderly made."[108] Although the colony was reluctant to allow demolishing of dams, it did believe action was needed for "the preservation of the fish," particularly because "by reason of the many dams erected . . . said fish are diverted in their passage, to the great decay and ruin of such fishery."[109] The colony passed legislation requiring "whosoever shall hereafter erect or build any dam across any such river or stream where the salmon, shad, alewives, or other fish usually pass . . . shall make a sufficient passage-way for the fish to pass up such river or stream through or round such dam, and shall keep it open for the free passage of the fish from the first day of April to the last day of May . . . and owners or dams erected . . . where fish can't conveniently pass over, shall make a sufficient way either around or through such dam [in the fall]."[110] The New England colonies also enacted statutes empowering the towns to restrict fishing and to limit fishing to certain days and certain hours in order to allow the fish free passage upstream.[111]

The coming of the American Revolution did not stop the flood of laws governing the protection of fish.[112] The conflict between fishers who wanted access to as many fish as possible, dam owners who wanted to keep their dams up the year round, and fish who wanted to complete their life cycle continued to plague the new state legislatures. Attempts to avoid laws protecting fish led to court challenges.[113] Where existing "laws already made are not sufficient to prevent the destruction of the fish," new, even more stringent laws calling for the destruction and "demolition" of obstructions were passed by the newly formed New England states.[114]

The flow of legislation concerning fish and fishways reflected a general assumption by the colonies and the states that fish were a vital resource that "furnished food to the inhabitants."[115] Even when mill owners received the right to flood meadows and fields, legislatures remained reluctant to allow activity that threatened the region's fish resources.[116]

These early obstructions that blocked the migration of fish, "which from time immemorial had passage up the river to cast their spawn," were usually small affairs.[117] The dams were a few feet high, with mill gates that could be raised to open the dams and allow the free flow of the streams and rivers. During the winter lumbering season, when the logs could be brought out of the woods over snow and frozen ground, local logs were dragged into the sawmills for cutting. Logs from farther away were floated to the mills with the first clearing of ice from the river. By the end of March, well before the migratory fish reached the dams upriver, the sawmills had finished their heavy cutting.[118] While gristmills operated throughout much of the year, the majority of the milling occurred during the fall and winter months.[119] Therefore, when the fish came upriver, gristmills, like sawmills, could lift their gates and allow the fish passage, or they could construct fishways to allow the passage of fish around the dams.[120]

The ability to lift the dam gates to accommodate migratory fish did not guarantee that the fish would be let through. In 1789, the towns of Stoughton and Sharon became concerned that mill dams on the Neponset River were cutting off the migration of shad and alewives. In response to a petition from the two towns, the Massachusetts legislature passed an ordinance requiring fishways around the dams or the opening of the dam gates during the fish migrations. In 1808, two mill owners, Edmond Baker and Daniel Vose, claimed the law requiring them to let the fish through "interfered with their right to enjoy their estate [running their mills at full capacity during the fish migration period]." Unfortunately for Baker and Vose, the court ruled that the right to build a dam implied limitations, and one of those limitations was "to protect the rights of the public to the fishery, so that the dam must be so constructed that the fish should not be interrupted in their passage up the river to case their spawn. Therefore every owner of a water mill or dam holds it on the condition, or perhaps under the limitation that a sufficient and reasonable passageway shall be allowed for the fish. This limitation being for the benefit of the public."[121] In 1813, the legislature passed statute 1813 c43, providing that if fish-ways were insufficient, "the dam may be removed or abated as a nuisance, in the same manner as other nuisances may, by law, be removed or abated."[122]

The early-nineteenth-century mill owners who had to deal with these fish acts and court cases were mostly operators of gristmills and sawmills. But even as the legislature was trying to protect fish, the nature of the mills themselves was changing.

Rivers and Canals: The Arteries of Trade

Neither market agriculture nor manufacturing could develop—no matter how scientific the farmers were or how efficient the manufacturing process was—without accessible markets. That meant convenient and cheap transportation.[123] New England's natural transportation arteries were its rivers. Goods leaving the region could float and sail down on flat boats with short sails, or on log rafts.[124] Incoming goods would sail on coastal or oceangoing vessels up to either the first fault along the river or the end of the larger tidal basin. Goods would then be reloaded onto smaller river sailing boats with shallow draws and short flat sails for the trips farther upstream. On the Connecticut, oceangoing vessels could bring incoming goods up to Middletown or Hartford. Between Hartford and the northern hill country, a series of fault lines cut across the river, causing falls and rapids. The first of these was the Enfield rapids. Between Enfield and the northern reach were three major falls—one at South Hadley, north of Springfield; one at Turners Falls (also called Millers Falls), north of Deerfield; and one at Bellows Falls, north of Brattleboro, Vermont. Farther north, there were Sumner's Falls at Ottauqueechee near Plainfield; and Olcott's Falls (also called Lebanon Falls) at West Lebanon, Vermont, and White Water Junction, New Hampshire.

Traders, merchants, and prosperous farmers north of each of these obstructions believed that if a system of dams, locks, and canals was built around the falls and rapids, goods would flow more quickly in and out of their stores, and their region would prosper. The dams to feed the locks and canals had to be significantly bigger and more complex than the simple earthen, stone, boulder, and log dams used to power a grindstone.

The capital and engineering required to accomplish this task were much more extensive than those needed to erect a gristmill and dam. Visions of extensive trade flowing in and out of the region and profits from operating these new locks and canals danced before the eyes of the region's elite.[125] These projects required capital, engineering knowledge, and support beyond the capacity of local merchants and prosperous farmers. Pushed by the project boosters, the various New England state legislatures incorporated canal-building corporations with power to raise capital, buy property, and flood fields and meadows under the terms of the Mill Acts. In some cases, they supported the companies by giving them the authority to raise extra capital through state-sanctioned lotteries. Using connections established through their trading activities, the organizers behind these projects tapped into capital in Europe as well as at home.

Construction for the first of these major dam-canal works on the Connecticut began in Vermont at Bellows Falls. Vermont chartered the Company for Rendering the Connecticut River Navigable by Bellows Falls in 1791, and work began in 1792. Capitalized primarily by money from London, the canal and lock system was under construction for ten years before the first boat passed through in 1802. By the time it opened, $105,338.13 had been spent, a sum far in excess of anything a local mill owner could afford.[126]

In February 1792, a group of wealthy Hampshire County merchants formed the Proprietors of the Locks and Canals on Connecticut River. They received a charter from the Massachusetts legislature to build canals around Hadley Falls and around Turners Falls. On March 21, with capital from Holland, they announced a meeting to choose officers, decide on shares, and make decisions about surveying and purchasing land for the canal at South Hadley.[127] By December 1792, the proprietors were seeking and would pay cash for 70,000 feet in "length of timber of yellow, white, or pitch pine, spruce, hemlock, chestnut, or oak in pieces of any length not less than 25 feet to be straight and of any size not less than one foot in diameter in the small end the whole of the above timber to be delivered at Boardway on the Connecticut River, South Hadley in the first week of June."[128] The canal system at South Hadley, when finally completed in 1795 at a cost of $81,000, had an eleven-foot-high dam and an inclined plane on which a carriage would move up and down, drawn by a waterwheel. Six years later, the inclined plane was replaced with an eight-lock canal two and a half miles long.[129] Two years after construction began at South Hadley, work began on the canal around Turners Falls. When finished, its dam was 17 feet high and 972 feet long. The canal was two miles long, and twenty feet wide. A second dam two miles farther upriver completed the system, which had a total of ten locks. As with the South Hadley project, the cost of this system was over $81,000 by the time boats began passing through the system and paying their tolls in 1795.[130]

In 1794, the Vermont legislature incorporated the Company for Rendering Connecticut River Navigable by Water-Quechee, and New Hampshire incorporated the same company in 1796 in order to build a canal around Sumner's Falls. The canal around these falls was short and required only two locks. Seven miles north of Sumner's Falls was Olcott's Falls or Lebanon Falls. Vermont created a corporation to begin work around these falls in 1795 and finally completed the dam, canal, and locks system in 1810 for $40,000. The last of the Connecticut River system dams, canals, and locks were built by the Connecticut River Company in 1828 around the falls at Enfield.[131] With the completion of the locks and canal system on the Connecticut, the flow of goods out of the region increased. By the 1830s, goods went downriver for two to three dollars a ton and upriver for between five and six dollars a ton.[132]

The dams built for the canals and locks dwarfed those built by millers. Like the system at South Hadley, they involved the expertise of engineers, lumber and stone imported from afar, the labor of hundreds of workers, and the pooling of capital in amounts unimagined by earlier generations of dam builders. Not only were the dams for canals bigger, but they flooded substantially more land. The incorporating language of these canal-building companies contained the right of these companies to seize land for digging the canals and their obligation to pay persons whose land was damaged. As landowners went to court to claim damages for nuisance against the dams, the legislature and the courts applied the language of the Mill Acts to these dams.[133]

Because canal dams were so much larger than traditional mill dams, the flooding they caused raised new problems. The dam at Hadley Falls raised by four feet the water behind it for ten miles, flooding extensive meadows and leaving stagnant pools of water where there had previously been flowing creeks and streams. The people of Northampton, who had once taken pride in living in "one of the healthiest towns in New England . . . after the erection of the dam were extremely afflicted with the fever and ague: a disease which . . . was unknown in this town for more than sixty years."[134] Faced with an increase of sickness and disease, the people of Northampton fought back.[135] Claiming that the dam was a nuisance under common law, a number of suits were initiated against its proprietors. The suits claimed that the dam caused illness and blocked the migration of fish.[136] In 1801, the legislature appointed a committee to go to Northampton and South Hadley and "inquire into sickness and the state of locks and canals and the state of fisheries."[137] The same year, the jury at the court of common pleas for one of these suits found for the plaintiffs and declared the dam a nuisance. The judge ordered the company to take away part of the dam and lower the water. The proprietors then faced the possibility of an endless number of suits, and they lacked the capital to rebuild the dam and canal in such a fashion as to not flood the fields of Northampton. In desperation, the proprietors turned to the state for help. In 1802, the state granted the proprietors a lottery grant, a legislative privilege that was used in the late eighteenth and early nineteenth centuries to raise money for public and private projects. The lottery "yielded them a considerable sum," which was used to rebuild the system, eliminating the inclined plane, deepening and lengthening the canal, and structuring the dam and locks in such a fashion as to avoid flooding as much of Northampton's fields as they had before.[138]

Lotteries were not the only way the major canal projects were supported. When the canal companies built dams in order to divert water into their canals, the dams also held back water, which could be sold as waterpower. In 1795, the State of Massachusetts added an additional act to the act incorporating the Middlesex Canal Company (statute 1798 c.6). This empowered the corporation to hold mill sites connected with the canal and to erect mills upon those sites.[139]

Mills and lumber. Connecticut River below Bellows Falls, Vermont.

"A Country Changing"

The day before he died on January 10, 1817, Timothy Dwight made plans for the publication of his journals. Dwight wanted to present his New England to the world. But he realized that even as he was traveling, New England was changing. "A country changing as rapidly as New England must, if truly exhibited, be described in a manner resembling that [with] which a painter would depict a cloud."[140] The most pronounced change that Dwight noted was the "conversion of a wilderness into a desirable residence for man." That conversion changed "a forest . . . within a short period into fruitful fields, covered with houses."[141] The forests of New England were not being cut only to make way for farms. They were also being cut to build the factories that were springing up beside the Connecticut and its various tributaries. Timothy Dwight noted that northern New England was "furnished with millstreams and millsites. Manufacturers are begun in various places, and ere long will be an object of primary attention to the inhabitants."[142] Dwight commented that along these millsites, manufacturing enterprises came, and around the mills came towns. "Common laborers, diggers of canals, lumber merchants, dealers in hardware, brass, and iron founders, burners of lime, carpenters, masons, carriers, wagoners, sellers of wood, and blacksmiths are all employed in greater or less degree by the erection of a cotton manufactory. To these are to be added the superintendents, clerks, overseers, agents at home and abroad, dyers, and that numerous class of men and women, and children who are immediately employed in the manufacturing of yarn. A manufactory of 1500 spindles will soon accumulate a population sufficient to form a village."[143]

The period from the late eighteenth to the mid-nineteenth century saw the shift from processing agricultural goods for local consumption, such as milling grain, cutting lumber, and tanning hides, to manufacturing goods (initially from local raw materials) for increasingly distant markets. With manufacturing activity, old gristmill and sawmill sites were converted to new mills with water-driven machinery, and rivers and streams were dammed to ever greater heights. New England's land, its resources, and its people effected this transformation and were in turn affected by it.

2

From Milling to Manufacturing

From Villages to Mill Towns

The new world of New England was one of factories and factory towns, as well as farms and forests. It was a world where farmers, looking to those factory towns for markets, plowed their fields deep and intensively managed their land. It was a world where lumbermen stripped mountainsides of their forest cover to meet the cities' growing appetite for lumber. It was a world of managed and controlled nature.[1] It was also a world of rapid change, and increasingly after 1800, the force behind that change was the coming of the manufacturing mills.

Levi Shepard's 1788 duck-cloth factory was of a different type than the traditional mills of New England. Although mills that spun or fulled cloth had long been part of rural New England, Levi Shepard had a different market in mind when he encouraged local farmers to bring him their flax. Shepard wanted to take material from the countryside and, with the help of "workers employed," "manufacture" it into a commodity for sale.[2] Shepard's decision to focus on manufacturing for distant markets represented a new world.[3]

Mills and Manufacturing

Manufacturing in rural New England began small. And although it made a huge impact on travelers such as Timothy Dwight, it grew out of, while at the same time it transformed, traditional rural society. The processing of goods of the countryside was an integral part of traditional New England life, whether in 1650 or 1800. In 1790, the *Hampshire Gazette* commented that although "a large quantity of woollen cloth are made in private families and brought to market in our trading towns, a great part of [the woollen cloth] is not calculated for market."[4]

The shift from milling produce for local use to manufacturing occurred initially for most of rural New England with the shift of small traders, merchants, and millers from processing for local farmers to processing for external markets.[5] Edmund Taylor of Williamsburg on the Mill River, for example, at the turn of the century added carding and picking machines to his gristmill. As he did for grain, Taylor processed the material from the countryside, keeping a portion of it as his pay. Once established, Taylor added a cotton textile factory to his enterprises, using the surplus water from his gristmill.[6]

The Bellows Falls Company of Vermont, besides running a country store, was using waterpower to run "wool carding machines" for local farmers. Like many other traditional millers and country store owners who dammed up a stream and put in a waterwheel to run a sawmill, gristmill, fulling mill, or even carding machines, the Bellows Falls Company processed produce for the local community. Customers brought to the mill products of their farm, which they had processed at the mill and then took home with them. Although the Bellows Falls Company may have kept some of the wool it carded and shipped it into the stream of commerce flowing out of the valley, the local community saw the mill's primary function as processing products for the valley's residents.[7]

In the early years of the nineteenth century, manufacturing in New England underwent a radical transformation and in doing so began the first stage of the long and complicated process of industrializing the New England countryside. When Thomas Jefferson imposed an embargo on goods coming into and going out of the United States through the War of 1812, imported manufactured goods became scarce in America. Shopkeepers and traders who found that they could no longer get finished cloth or paper began to look at the old gristmill and sawmill sites with new eyes. In 1808, Job Cotton took over Joseph Burnell's sawmill on the Mill River and brought in cotton textile spinning machines. On the Chicopee River, the Chapin brothers—William, Levi, and Joseph—built a small mill that by 1810 had two carding machines, a drawing frame, and two short spinning frames "of very rude construction." The Chapins employed eight to ten workers, who produced yarn that was put out to local women to weave. The workers had no regular payday. The company exchanged yarn and cloth at the local stores, and those it employed had credit at the stores.[8] By the time war broke out in 1812, old mill sites throughout New England were being converted into textile mills.[9] Anticipating increased demand for manufactured woolens, Stephen Cook in 1815 replaced his gristmill with a woolen manufactory, and Isaac Biglow made plans to build a cotton factory of five thousand spindles beside his gristmill.[10]

Peace with England in 1815 brought not only an end of war and of war orders for cotton and wool but also a flood of English manufactured goods to the American market. Most of the small cotton mills went under in the ensuing years. Job Cotton's mill closed its doors and stood idle, and Isaac Biglow never used his cotton factory as intended but outfitted it instead to make paper.[11] The years following the peace brought the first real depression to inland New England since Shays's rebellion. The depression may have closed Job Cotton's mill, but it did not return the valley to its more traditional past.

Although the general depression in the postwar years wiped out most of the small manufacturing operations, the better-capitalized mills with big dams, overshot waterwheels, and newer machinery weathered the depression and continued to grow. The new large dams and multistory factory buildings that utilized heavy, powered machinery to turn out woven cotton and woolen cloth needed wooden beams and flooring, machines, belts, and pulleys, candles and paper.

As New England began to manufacture goods for far-off markets, the very manufacturing process itself began consuming massive amounts of the region's resources and generated more manufacturing. Manufacturing needed machinery, and increasingly that machinery was locally produced. These mills not only gathered up massive amounts of waterpower and resources, they also gathered up hundreds and even thousands of people into new industrial cities, taking in food and water and disposing of waste.

Manufacturing along the Connecticut River Valley

At the turn of the nineteenth century, Timothy Dwight noted that mill streams everywhere were giving rise to an "unusual number of works erected for various manufacturing purposes, powder, paper, glass, etc."[12] Another Dwight, Theodore, noted in 1830 that indeed New England's gristmills had given over "within a few years [to] manufactures." "On the road up the Connecticut River . . . everyone must be struck with the size and number of the manufacturers which have been multiplied and magnified to such an extent all over the country."[13]

Most of these early manufacturing mills were located on small rivers and tributaries. Over the century, as engineering skill improved, technology developed, and more capital became available, manufacturers had the ability to capture more power from the larger rivers. As they did so, they moved to the falls of water on the region's great rivers. Around these falls grew industrial towns, then cities. These cities in turn reached out into the countryside, pulling more and more of it into their grasp. Included in the resources needed was water. And the Connecticut River had vast amounts of water.

Manufacturing along the southern Connecticut River Valley began early. Already by 1830, Middletown's "various manufacturers carried on with success," while Hartford had thriving paper, iron, and brass works as well as an established armaments industry. In the nineteenth century, tinware producers in Berlin, Connecticut, expanded to include brassware and hardware, and this metal-working industry spread to towns up and down the valleys between Middletown and Hartford.[14] On the west side of the river at Windsor Locks, paper mills and silk mills were built, while at Enfield, carpet making became a major industry.[15]

By the nineteenth century, Springfield's industry, led by the U.S. Armory, grew to be a crucial part of the town's activity. G. and C. Merriam bought the copyright to Noah Webster's dictionary and began to manufacture the nation's best-known reference book. Across the Connecticut River in West Springfield, along the tributary Westfield River, large-scale tannery factories transformed tanning into a huge enterprise consuming massive amounts of tree bark. The Clark Carriage and Wagon shops and several new cotton and woolen factories in Agawam village all employed large numbers of hands. The region's expanded manufacturing demanded factories and homes. New circular sawmills were built to cut lumber coming downriver. Unlike the older sawmills, these new mills were large-scale operations with dozens of workers constantly at work cutting and sawing. North of Springfield on the east side of the Connecticut, a number of manufacturing enterprises rose up along its tributary the Chicopee River. Ironworks triggered Chicopee's rise as a manufacturing center. During the early years of the century, a number of small cotton mills opened for carding and spinning cotton, although most did not survive the depression of 1816. In 1825, through the intervention of Edmund Dwight, Boston's textile investors (known to historians as the Boston Associates) began to take an interest in the waterpower of the Chicopee River.[16] The Chicopee was dammed, and the first of the huge multistory textile mills and attendant boardinghouses were built. More mills soon followed, so that by the end of the 1830s, Chicopee had four major textile manufacturing companies that employed hundreds of workers.[17] Edmund Dwight also persuaded Nathan Ames to move his edge-tool manufacturing company to Chicopee. Ames's company prospered, and in 1845, with his connections to Dwight, Ames bought up the local machine shops and began to produce textile machines as well.[18] Upriver in Palmer, the Thorndike Manufacturing Company, again the product of Boston investment, in 1837 built a gigantic stone mill that soon employed 450 workers producing cotton ticks, denims, and stripes.[19]

At Bellows Falls, Vermont, Bill Blake obtained waterpower rights in 1802 from the Company Rendering the Connecticut River Navigable by Bellows Falls and built a paper mill there, rebuilding it in 1812. He built still another paper mill in 1814 in Wells River, Vermont. Blake's mills proved successful, and soon he had teams throughout Vermont and New Hampshire buying up rags for his paper mills, which were turning out paper for sale around the country.[20]

On the outskirts of Springfield, John Ames built an extensive paper-making factory in the early 1820s that employed dozens of local workers to cut and sort rags and work the new multistage, water-powered paper-making machines. In 1827, another Ames, David, took over a small one-vat, three-hand, screw-press paper company in Chicopee and expanded it into a major manufacturing enterprise, introducing power-driven paper-making machinery and exporting paper out of the region. By the 1840s, towns up and down the valley were turning out paper for books, newspapers, stationery, envelopes, wrapping, and collars.

Factories grew up along tributaries to the Connecticut. In the 1830s, wing dams on either side of the river north of Springfield at Hadley Falls provided power for a series of traditional mills, along with the newer cotton and woolen mills. With an initial investment of $50,000, a group of Springfield and Enfield investors formed the Hadley Falls Company in 1827 in order to build a large cotton textile mill powered by water brought in by a wing dam. By 1832, the company was running 2,700 spindles and employing forty-six women, "daughters of local farmers," and twenty-three men.[21]

North of the falls at Hadley, the tributary Mill River had a number of silk manufacturing companies, which briefly encouraged local farmers to grow mulberry trees to feed the silk worms.[22] Farther north, cotton textile and woolen factories sprang up at Bellows Falls, along with Blake's paper mills, while at Lebanon, New Hampshire, woodworking shops and factories produced tools, furniture, and agricultural implements from lumber harvested from local forests. Between 1810 and 1850, manufacturing mills, machine shops, and woodworks multiplied along the rivers and streams running into the Connecticut.[23] At Franconia, New Hampshire, a significant ironworks developed.[24] By the middle of the nineteenth century, blacksmith shops and iron foundries had shifted to waterpower for operating bellows and, in the foundries, for milling iron casting. With this shift, more ironworks were located on waterpower sites.

The expansion of sheep raising in Vermont and New Hampshire provided wool for the expanding woolen factories sprouting up around the falls of almost every northern New England river.[25] At Hartford, Vermont, Albert Dewey dammed the falls of the Quechee River and began woolen manufacturing, producing 450 yards daily.[26] Dewey's factory on the Quechee was only one of many that sprang up in northern New England in the thirty years after the War of 1812 to take wool from the region's farmers.[27] In 1811, Colonel James Shepherd opened a small woolen factory on the Mills River, where spinning and weaving was done by hand. By 1818, Shepherd had expanded and installed power spinning mules and weaving machines. By the end of the 1820s, the company was converting fleece from Shepherd's and his neighbors' large flocks of sheep into broadcloth, which was then sold throughout the country.[28]

Factories and Farms

Factories transformed the land around them from farms and fields to other factories, tenements, roads, and canals. As they produced goods for export, they also became import centers themselves, not only of food and lumber but also of people. The mills and their machines needed operatives, who came from many of those same hills that furnished the lumber for the mills.

These manufacturing and commercial centers were surrounded by commercial farmers, while poor country cousins abandoned or scraped by on overworked hill-country farms. Where once there were villages with teachers, doctors, ministers, traders, merchants, artisans, clerks, and lawyers, all with their own gardens, cows, pigs, and chickens, now there were cities with hundreds of operatives and tradespeople, very few of whom could afford the time or had the space to raise their own food. The food that fed the growing cities of New England came from the region's farms, but these were very different farms from those of the eighteenth century. New England's cities took the farm produce from farmers who specialized in feeding urban people.[29] The milk, cheese, butter, and vegetables that poured into the cities to feed the operatives came from farmers who specialized in dairy products or commercial garden vegetables.[30] In 1846, for example, Jesse Chickering noted one of the changes that had occurred in Massachusetts agriculture: The "dense state of the population in the villages" had given rise to an "increase . . . of vegetables raised, such as potatoes, apples for eating, garden vegetables, and fruit."[31] New England farmers had always raised fruits and vegetables, but increasingly, nineteenth-century farmers focused their activity on perishable market produce, and the markets for that produce were local.[32] In a midcentury promotional bulletin, the Hadley Falls Company noted that "the position of a city in the midst of the splendid farms and fruitful meadows of the Connecticut Valley would be highly favorable for obtaining supplies; and the existence of such a market for their produce would be greatly beneficial to the interests, and stimulating to the industry, of the people of that part of the state engaged in agriculture."[33] As the *Hampshire Gazette* recognized in 1831, farms around the new industrial centers prospered when they shifted to sending food to feed the workers gathered to work in the mills.

"Go view the manufacturing villages," it urged, "go and view the farms which surround them ... see the universal prosperity which reigns there."[34]

The significance of this new market for agricultural goods encouraged the Palmer (textile) Company in 1831 to argue before the Massachusetts Common Court when its dam flooded the lands of farmer Isaac Ferrill that it should not have to pay Ferrill damages. The Palmer Company argued that because of the mills, Ferrill's land "could be sold more readily and at a higher price, and all the products of the tract command a readier market or higher prices."[35]

The growing market for perishable farm products did not ease the plight of the valley's hill-country farmers. The lowland farms on the more fertile soil were better positioned to take advantage of this market. They were also better equipped to use more modern agricultural tools and technology. The rocky, hilly topography, thin soil, and greater erosion of the hill farms meant that they were less capable of taking advantage of the new farming methods or the new local industrial city markets located on the riverbanks of the lowland valleys. Increasingly after 1830, as lowland farmers moved to intensive, market-oriented agriculture, hill-country farmers abandoned their farms entirely or began to harvest wood instead of crops.[36]

Forests and Factories

In addition to the produce from the surrounding countryside, the factory towns also consumed acres and acres of the region's forests. Hills and mountains far up the valley yielded up their forests so that dams, mills, and tenements could be built and heated, and tools and furniture could be fabricated.[37] To encourage manufacturers to build mills at their dam site, the Hadley Falls Company used a report detailing the wealth of available lumber from the valley's northern forests. The company noted in its promotional report that "there is not likely to be any lack of building materials at Holyoke. Lumber is brought greater or lesser distances down the Connecticut River from the forests on its banks."[38]

As the mills and mill towns grew in number and size, lumber floated in ever increasing volume down from the hills and mountains of Vermont and New Hampshire. The combination of burning and cutting forest land to clear farms and cutting timber for construction and fuel meant the baring of northern New England's valleys, hills, and mountains. Between 1820 and 1860, Massachusetts and Vermont went from over 60 percent forest covered to just over 40 percent, while New Hampshire went from over 60 percent to less than 50 percent.[39]

Lumber for building the growing manufacturing cities' factories and boardinghouses was only one of the demands on the region's forests. Wood for fuel to warm those homes and to power a new form of transportation penetrating into central New England also took its toll.[40] The railroads not only linked New Englanders together, but they also brought cheap Midwestern grains, beef, pork, and wool into the area, further eroding the position of northern farmers. Although the railroads undercut the hill-country farmers' market for traditional produce, they offered another market themselves. Railroads consumed massive amounts of wood for ties and fuel. In 1846, G. B. Emerson in his report to the State of Massachusetts on trees and shrubs of the state estimated that the 560 miles of railroad in Massachusetts were consuming 53,710 cords of wood annually.[41] By the end of the decade, the region had more than tripled its miles of railroad, and wood consumption reached hundreds of thousands of cords a year.[42] Farmers pressed by cheap Midwestern produce brought in by train now harvested the hills of the north of their forest cover to help feed the appetites of the new iron horses.[43]

Increasingly after the 1870s, New England's railroads turned to coal for fuel, and after midcentury, homes and industry in the region's cities did so too. The burning of coal eased the demand for cordwood while it increased air pollution. But the burning of coal did not end the pressure on the region's forests. It was not until the 1870s that New Hampshire's forest area stopped declining, while Vermont continued to lose forests until 1880, when only 35 percent of the state was forested.[44]

Timber harvests from New England's northern forests not only stripped the land of old-growth trees, but as George Perkins Marsh warned the Vermont legislature in 1857, the "clearing of the woods" would make "brooks and rivulets, which once flowed with a clear, gentle and equable stream . . . dry or nearly so in the summer, but turbid with mud and swollen . . . after heavy rains."[45] The Massachusetts Board of Health noted fifteen years later that surrounding woodlands protect the banks of lakes and ponds from washing away and prevent soil from being carried into the lakes by rain: "As a country is denuded of wood, it becomes parched and arid."[46]

Despite the warnings of Marsh and the Massachusetts Board of Health, lumber harvests continued, and logs increasingly became the main resource moving downriver. Each winter, hundreds of thousands of logs were cut out of northern forests and pulled by oxen to the Ammonoosuc River, the Moose River, and the White River to be floated down to waiting mills farther south.[47] The continued demand for lumber for the factories and tenements of the industrial centers growing up at Holyoke, Chicopee, Springfield, and farther south at Hartford sent more and more woodsmen out into the winter forests to bring down trees and send them south. The dam that went up at Holyoke at midcentury interrupted but did not stop the flow of lumber. The Holyoke Water Power Company noted that throughout the second half of the nineteenth century, spring freshets brought down millions of feet of logs from the northern reaches of the Connecticut River system to be cut by the Holyoke Water Power Company's subsidiary sawmill, the Holyoke Lumber Company, and "used in building the Holyoke factories and homes" or "sent on the way . . . to Hartford." In 1876, for example, the Holyoke Lumber Company sent more than six million feet of logs downriver to Hartford.[48]

The railroads increased the access of dairy and potato farmers to the coastal cities. Cheese, milk, butter, and potatoes joined cordwood as products leaving the northern valley for urban markets to the south.[49] But the railroads also brought competition. The small manufacturing villages of New England supplied local markets with furniture and tools. Now, local manufacturers could no longer compete with manufacturers who produced goods on a larger scale. After 1850, lumber was increasingly sent downriver to be milled in the large mill factories such as the one in Holyoke, while furniture and tools were made elsewhere. Before 1850, more than half the woolen mills were located in rural townships of less than five thousand people, such as Dudley or Oxford, taking in local wool for manufacturing, but after midcentury, these mills could no longer compete with the larger ones at Lawrence, Holyoke, or Worcester.[50]

Holyoke: The Epitome of a New England Mill Town

On July 24, 1882, Theodore Lyman—a wealthy Boston Brahmin, Civil War veteran, Harvard Overseer, amateur scientist, Massachusetts commissioner of fisheries, and independent Mugwump candidate for Congress—in answer to a question about protective tariffs, quoted his father, who had been mayor of Boston in the 1830s. "My father use to say very truly as long as New England was overlaid by ten feet of gravel, she would have to manufacture or starve." Lyman went on to say that New England had "an extraordinary amount of manufactures and of the greatest variety. . . . I can hardly believe my eyes such is the enormous growth." "What would Massachusetts be like" without its industry? Lyman wondered. "You would see no mills and no dams, nothing but a few grist mills here and there and houses whose occupants raised such crops as they could from the scanty soil."[51] Of course, before the massive manufacturing establishments arrived, Massachusetts had both dams and mills, but by the 1880s, New Englanders could scarcely imagine a world without industrial enterprises; when they did, some of them viewed it as empty and miserly. It was not the mills of Levi Shepard or Isaac Biglow or Job Cotton that Lyman found so "extraordinary." It was the Boston Associates' large, heavily capitalized mills that impressed Lyman.

The success of the Boston Associates' investment in Chicopee and the completion of the rail system to western Massachusetts and to Albany, New York, raised interest among investors in developing the waterpower at Hadley Falls into an industrial complex such as the ones in Manchester, and Nashua, New Hampshire, Lowell, or Lawrence, Massachusetts. These investors noted that by the 1840s, although "every fall in the rivers of the second class, such as the Merrimack and Charles, and, indeed, every 'privilege' upon the smallest and most insignificant streams, has been seized upon for the use of some mill or factory, the noble Connecticut has been long almost entirely neglected. . . . This neglect would be inexplicable, were it not obvious that the extraordinary magnitude of the enterprise required, in corresponding degree, more than ordinary forethought, skill, and capital, for its accomplishment."[52] These men had successfully dammed the Charles and Merrimack, and their mills on those rivers produced unimagined profits.[53] Damming the Connecticut at Hadley Falls represented a bigger, costlier, and more challenging engineering project, but the Boston investors had the capital. They also believed they could muster the engineering capacity to dam the river and control its nature.[54]

In April of 1847, the Hadley Falls Company agreed to sell its mill, property, water rights, and name to Edmund Dwight, who represented the interests of the new investors.[55] Getting all the land was not easy. When the company agent went to buy Samuel Ely's land, he was met with hostility. Ely, "an old-fashioned farmer," "didn't want to see the corporations control everything, and he was sorry they had come." Ely made his dislike clear when he shot at the company agent who came to try and persuade him to sell. By the fall of 1847, the Boston investors had spent some $300,000 for land and water rights, including the land and water rights of the old Proprietors of Locks and Canals on the Connecticut River, for a site with the greatest potential for mill development and waterpower in New England.[56] The company estimated that with a flow of 6,980 cubic feet per second in the driest month of the year, and with a fall of 59.9 feet, the river would deliver 30,000 horsepower, or 550 mill powers.[57] Based on their experience at Lowell, the Holyoke investors estimated that "a first class manufacturing requires several mill powers." Holyoke, they assumed, had mill power sufficient for a hundred establishments of average size. If just half the potential was developed, there could be more than a million spindles, employing 200,000 people.[58] The company began surveying and developing plans for a dam, a series of canals and races, and sites initially for six mills, mechanic shops, boardinghouses, a fish way, a transportation canal, and rail lines.[59]

By November 1848, the dam was complete. As water began to build up behind the completed dam, the unwarranted nature of the original investors' optimism soon became apparent. Within three hours, the dam was "gone to hell by way of Willimansett" at a loss of some $38,000 and a massive amount of northern forest hemlock.[60]

Not discouraged, the investors immediately hired a new engineer, John Chase, to design a new dam. Construction on the new dam began in April 1849 and was completed by October 22, 1849. The new dam was thirty feet high and one thousand feet long, significantly thicker than the first dam at ninety-two feet deep, and built with interlaced solid timber (four million feet, with the smallest log twelve inches in diameter at the small end), with stones packed into the open spaces. The top was covered with sheets of boiler plate. This time, when the water was released behind the dam, it held. Indeed, the dam held for another fifty years before it was replaced with a newer steel and concrete dam in 1904.[61]

With the successful completion of the dam, the Hadley Falls Company built boardinghouses to house the operatives. The first cotton mill was built, and spindles began spinning in 1850. By 1852, the first mill ran at full capacity, and a second began operation. In 1853, the company had two mills, each 268 feet long, 68 feet wide, and five stories high. The

Holyoke Dam, Holyoke, Massachusetts.

From Milling to Manufacturing

boardinghouses for the operatives were "built of brick, in the most substantial style, and are supplied with all the usual conveniences of modern dwelling-houses."[62] In 1854, a third mill, the Hampden Mill, began operation. The town of Holyoke began to fill with operatives, mechanics, clerks, and tradespeople.[63] The boardinghouses quickly filled with workers.[64] By 1860, 4,632 people lived where fifteen years earlier the homes of only a few hundred farmers and a few operatives were scattered about.

Despite better times and growth, in the late 1850s, Holyoke's investors continued to have trouble. The investors in Holyoke built more than textile mills. They also built the dam and canal system to supply waterpower, operated machine shops to supply and repair the mills' machinery, created mill sites, and built and managed the boardinghouses. The huge expenditures required to build the Holyoke project burdened the company with debt it could not liquidate. In 1859, the Hadley Falls Company went into receivership and was auctioned off. The new owner reorganized the corporation into the Holyoke Water Power Company.[65]

Conditions improved in Holyoke with the increased demand for textiles. This meant more employment at the two Lyman Mills and the Hampden Mill. In 1859, the company built an extensive gasworks to convert coal into gas to light the new mills, which operated long into the dark winter evenings. By 1881, the gasworks was producing twenty-seven million cubic feet of gas for the city's residents and businesses.[66] Although Holyoke was beginning to realize the physical dream of its original investors, if not the financial one, large numbers of mill sites remained empty and potential waterpower unused.

When the original investors began their Holyoke project, they assumed that the mill sites would go to textile mills, as they had at Lowell and Lawrence. But by the end of the 1850s, the optimism about the market for textiles began to wane. Investors, including those in Holyoke, were reluctant to put capital into new textile mills. The Holyoke Water Power Company began to look beyond textiles in an effort to sell mill sites and waterpower. Paper manufacturing, an industry already established in the valley, was an obvious candidate for the mill sites. The Parsons Paper Company, which developed from a flour mill site in Holyoke, expanded to become the nation's largest writing and envelope paper producer.[67] The Parsons's success paved the way for other paper companies. By the 1860s, the Holyoke Water Power Company was eagerly selling sites to paper companies. During the early seventies, paper-mill building increased so that in 1874, there were eight paper companies in Holyoke employing 1,620 workers, four paper-collar companies employing a total of 215 workers, and two book and news companies employing 75 people.[68] By 1880, papermaking, with twenty-three mills producing 150 tons of paper daily, had become the city of Holyoke's most important industry. Holyoke had become "Paper City." To process rags into stuff, (the mix that gets pressed into paper) paper manufacturing required pure water, and with the development of the large Fourdrinier machines, it also demanded extensive power.[69]

To its paper factories, machine shops, wire mills, and textile mills, expanding Holyoke added thread factories, a silk factory, several woolen mills, and a worsted mill. The mills, along with boardinghouses, stores, and a hotel, were built of brick, and to make those bricks, a brickworks was established that by the middle of the 1870s employed three hundred men who turned out ten million bricks a year.[70]

The increased demand on the company's waterpower capacity, particularly by the new paper mills with their rapacious new machinery, led the Holyoke Water Power Company to develop a means of more accurately measuring and charging for waterpower use by its customers. Since the original owners had based their concept of waterpower use (waterpower equaling the power to run 3,584 spindles sixteen hours a day) on the millpower unit of measurement, around the waterpower used by textile mills, they had no way of knowing how many mill powers the new individual waterwheels were using.[71] In 1881, they developed a gauge that was placed on each waterwheel to measure the waterpower used by the mill owners.

By the second half of the nineteenth century, Holyoke had become the epitome of a New England industrial town. The city had become one of the towns Theodore Lyman believed were the salvation of New England for their replacement of farms with industry, but one for which he cared little—an irony, considering that his uncle and father had been early investors. "How I detest manufacturing towns" he complained, "with their humming machinery, their prosaic faces, their long hard streets, ugly brick buildings and sickly smells. It is strange that this foundation of national power [and, it should be noted, of personal wealth] should also have so much that is deadening to the soul."[72]

As Dwight predicted almost a half century earlier, waterpower brought mills, which brought people.[73] Around the Holyoke mills, a population did gather of operatives, clerks, overseers, and laborers. They moved into the boardinghouses and other residences. There they ate, washed, and excreted waste.

Despite the original investors' claims that "Holyoke is so situated as to be as well favored by nature as any [town]," and that the town was "adapted to convenient and healthy residences," nature was also formed, shaped, and controlled by the very process of damming the river and building canals, mills, boardinghouses, shops, streets, and railroads.[74] When Holyoke was the village of Ireland Parish of fewer than two thousand inhabitants scattered about a rural and semirural countryside, it was indeed a healthy town "favored by nature." Its residents drew their water from streams, springs, and wells. They gave their food scraps to their pigs and had outhouses for their waste. They threw their wash water and slops out onto the yard to be absorbed into the ground. Even in the industrial village along the river, the shops employed so few workers that outhouses, privies, and a running ditch to the river easily accommodated the waste and sewage. But the development of large-scale manufacturing enterprises, and the communities of operatives that gathered around those enterprises, changed the environment in which people lived, prospered, sickened, and died. Outhouses and throwing wastewater out the back door no longer sufficed. Waste sewage flowed through ditches to the nearest outlet to the river, where it mingled with the chemicals, bleaches, dyes, and washes of the expanding textile and paper factories.[75] The Massachusetts State Board of Health noted in 1871 that "inland streams which had initially been dammed for rural sawmills and gristmills are by and by turned to other uses. Wherever a good water privilege exists there soon spring up various manufacturers, requiring many laborers and the greatest economy of the power of the stream. The village becomes a factory town, . . . the dam is soon raised, and where instead of a free movement of the water, tumbling over the fall in its superabundance as in former days, we find a great reservoir."[76] Around these reservoirs were factories that created both the reservoirs and the surrounding towns and cities. These newer communities were also increasingly unhealthy places to live.

Holyoke: A Place to Live

Life in Holyoke was constrained by both nature and the nature of the city. The work was hard and long. The world of Holyoke outside of work was ugly and polluted. As with many other New England industrial cities, the falls of a river had given rise to a huge industrial enterprise. This industry clustered together hundreds of people and overlaid the rural countryside with factories and tenements, shops, streets, and canals. Walking down Holyoke's Fountain Street, a nineteenth-century visitor would have a hard time distinguishing it from a similar street in Lowell, Lawrence, Chicopee, Fall River, or Manchester, New Hampshire. The factories, the canals, the streets, the crowded tenements and boardinghouses, and the mass of people entering factory gates in the mornings and exiting in the evenings became the common forms and images of the New England industrial city.

As the factories were built, people flocked to Holyoke in hopes of finding jobs in those mills. By 1870, Holyoke had more than doubled its 1860 population to over 10,000. In 1880, 21,961 people lived there. Indeed, within Holyoke in just two decades, the small pristine village of Ireland

Parish had become an "overcrowded, dirty," smoke- and dust-filled, "foul smelling" industrial setting.[77] Although Holyoke offered employment and housing, life in the city was also "deadening." Accidents in the mills were frequent. Workers were at the mills thirteen and a half hours a day with only two half-hour breaks. And wages for these many hours were low.

Although the company advertised its boardinghouses as "being supplied with all the usual conveniences of modern dwelling-houses," conditions were not comfortable. The strictly supervised boarders lived in cramped, overcrowded rooms.[78] Despite the water piped in from the company reservoir, until the 1860s, the boardinghouses relied on outhouses or privies more appropriate to scattered country homes than to highly built-up tenements with dozens of inhabitants.[79] The few sewers that the company built flowed into the canals.[80] Complaints about the stench finally led to the building of a sewer in 1857 that diverted sewage away from the canal. By the river, an Irish shantytown referred to as the Patch grew up with neither a clean water supply nor a sewer system. Wastes from the Patch ran into the river near the intake pipes that took in water for the reservoir.[81]

In 1863, the city's mortality rate of thirty-three per thousand was one of the highest in the region. Sewer lines were not extended to the Patch until the mid-1870s, and the tenement houses that the Hadley Falls Company had been so proud of in 1853 were twenty years later sanitary disasters. The Massachusetts Bureau of Labor Statistics reported in 1875 that "the sanitary arrangements [in Holyoke] are very imperfect.... Portions of yards are covered with filth and green slime, and within twenty feet, people are living in basements." The shocked health investigators noted that because of the "unhealthy" conditions, "it is no wonder that the death rate in 1872 was greater in Holyoke than in any large town in Massachusetts, excepting Fall River, and if an epidemic should visit them now, in the state they are in, its ravages would be great."[82] Even without an epidemic, health conditions in Holyoke were poor. Diphtheria, dysentery, typhoid fever, measles, scarlet fever, TB, and cholera took their toll on the growing city's population.

The 1871 drought and the breakdown of the pumps directing water into the reservoir convinced the town to develop a source of pure water. In 1873, the town bought land and two lakes three and a half miles west of the city, two mountain streams, storage reservoirs, and the southwest branch of the Manhan River and began pumping pure water to the city. By the end of the year, more than two-thirds of the city's families had access to safe drinking water.[83] The city of Springfield confronted the problem of clean water twenty-five years earlier than did Holyoke. While the Boston investors' engineers were working on their first unsuccessful dam at Holyoke, Springfield's leaders got the state legislature to pass legislation, statute 1848 c303, to create an aqueduct corporation to bring water from the springs feeding the Town Brook to Springfield to supply the city with "pure water." In a court case concerning that system, the court ruled that "the supply of a large number of inhabitants with pure water is a public good." Access to clean water did not answer all of Holyoke's health problems. The new water system and an expanded sewer system lowered the city's mortality rate to 21.67 per 1,000, but that was still far greater than the 15.3 in Springfield. Although clean water dramatically lowered the city's mortality rate, a third of the city's residents were still drinking questionable water.[84]

Holyoke, like Springfield and Chicopee, worked to bring its citizens water uncontaminated by its own sewage.[85] As the Massachusetts State Board of Health realized in 1872, "when people are massed together in crowded towns the local supplies of water inevitably fail, and it must be finally brought to them from without, and having been so brought and used, it must be carried away again."[86] In dumping its wastes into the Connecticut River, Holyoke extended its reach downstream, as well as upstream. The city reached out into the countryside to pull in clean water, as well as food and lumber, for its inhabitants, while it touched the downstream countryside with its industrial and human wastes. The carrying away of the wastes of the city and its citizens created a condition that would continue to plague New Englanders.

3

Cities and Industry, Sewage and Waste

In 1886, James Olcott, a farmer, "having been bred in the old anti-slavery reform," gave a speech before the Agricultural Board of Connecticut. Recalling an earlier age, he encouraged his audience and "the common people" of Connecticut to "agitate, agitate," in order to "cleanse" the state of the "social evil" of the pollution "by sewage from families and factories, festering in every pool, and mill pond—formerly trout holes." Olcott reminded the farmers that "our best hold on polluted streams reform lies in the fact that the mischief has brought on us its calamitous consequences in this country with such rapidity that men and women too not very grey-haired and in full bodily and mental vigor can shut their eyes and review the whole matter from its beginning."[1] The history Olcott conjured up was the transformation of a clean, clear environment from "one of the most salubrious to one of the worst in the world." The change was intimately linked to the rise of industrial cities like Bellows Falls, Chicopee, Hartford, New Britain, and Holyoke.[2] Although Olcott's remembrance of the past was partly colored by romantic notions of a purer age, the pollution he pointed to was indeed a problem of growing obviousness and concern.

Reflecting the rapid change that had occurred over the last quarter century, the Massachusetts State Board of Health complained that with the growth of densely populated industrial cities, the old habits of disposing of waste contributed to "a large part of the filth in our state," and that "often the water which is used for domestic purposes [is disposed of] by being thrown upon the surface of the ground, or collected in loose-walled vaults and cesspools," which might have been acceptable in a rural community but caused concern in the new industrial cities.[3] As the New Hampshire Board of Health noted in 1887, looking back over the last few decades, "when men mass, . . . the conditions at once become aggravated. . . . Man comes in with his artificial constructions and sweeps away much of this economy of nature."[4] For the New Hampshire Board of Health, nature, like a family, had an economy made up of many interdependent members. Nineteenth-century thinkers viewed an economy as a system of interdependence.[5] Because of the works of Adam Smith and his followers, by the late nineteenth century, the economy was also seen as a naturally regulating system governed by the invisible hand of supply and demand. In a world of farms and small villages, nature indeed seemed to function as a self-regulating system of supply and demand. A farmer removed crops from the land to feed his family and animals but returned to the land, as fertilizer, the wastes of those animals in the form of manure. In such a world, wastes were seen not as a problem but as a resource to be recycled back into the soil. As long as nature was in balance, human and animal excrement would not be wastes so much as components of an interconnected system. With the construction of industrial cities and the massing of humans in them, people seemed to have disrupted the natural balance. Nature's invisible hand no longer seemed to function.[6] Wastes that for the rural farmer had been mostly animal manure and spread fertilizer were now primarily human excrement and organic and chemical industrial refuse discarded into streams and rivers as problems and called pollutants.

People and Their Wastes

In 1810, William, Levi, and Joseph Chapin opened a cotton mill on the Chicopee River. The mill spun yarn, which was given out to local women to weave. For the Chapins, a privy not much larger than the ones by the homes their workers left in the morning to go to work easily met the needs of their "six or eight hands."[7] The new mills that grew up in the expanding industrial cities along the rivers of New England were not only different in their organization of production and capturing of power but also in their drawing people together on a scale unimagined by the Chapins.

People had been massed together in New England before the coming of the mills, but only in the cities of the coast, particularly Boston. The Massachusetts capital was nestled in a tidal bay and drained on the west and north sides by tidal rivers, and the people of Boston lived and worked in a variety of locations scattered about their urban space. This was not the case for the new inland industrial cities, which were located on freshwater streams that flowed downriver through other communities before emptying into the sea.

The residents of these new industrial cities were not scattered about the urban landscape in a variety of small workshops, stores, and warehouses but concentrated during working hours in large factory buildings. The new five-story mills crowded hundreds of workers together. In 1870, the two Lyman Mills at Holyoke employed more than eleven hundred workers in a space with a total ground floor of 36,448 square feet. A New Hampshire doctor, A. H. Crosby, estimated the average person's daily amount of solid and fluid excretal sewage in the nineteenth century to "equal about .03 cubic feet per day per person with another 3 cubic feet per person per day of water from water usage."[8] The human sewage of that many people working together for thirteen and a half hours a day was immense.

Companies usually built large privies or water closets behind the mills. These emptied into ditches, which then ran either into sewer lines or directly into the nearest stream. Many mills built their privies or water closets over streams. In Hartford, the Connecticut State Board of Health, while investigating a high prevalence of typhoid fever among the operatives of a large factory, found that the privy serving the three hundred mill workers emptied into a trough that flowed into a stream supplying the mill's drinking water. When there was sufficient rain, the stream washed the sewage away, but during the summer, the stream was "offensive."[9] At another mill site, the health inspectors found a "foul privy" over a long shallow trough that discharged into a stream. At a third mill, a privy at the back of the building for the mill's two hundred men "had a shallow trough two feet deep and wide. . . . The trough was flushed by rain water . . . [and] was connected to two large cesspools . . . [and] the overflow emptied into another ditch."[10]

The companies that built the mills had no model for what to do with the human waste of their employees other than that of the homes and shops of the small manufacturing villages, or at best that of the large commercial cities of the coast. Unfortunately, neither model was particularly adequate for the new industrial cities. For small manufacturing villages, privies emptied into ditches and cesspools and could accommodate small numbers of workers without necessarily endangering the health of the community, if the drinking water sources were far enough separate from the ditches and cesspools. When sewage became a major problem for the larger coastal cities with access to tidal flow, as it did in Boston in the 1830s and 1840s, they built sewer lines draining into tidal waters.[11] Because the tidal water was brackish and undrinkable, the larger towns' drinking water supply came from wells or was piped in from distant ponds or streams.[12]

When corporations began building huge multistory mills, the builders did not think human waste would be a problem. Nineteenth-century health reformers complained that "the neglect of proper provisions to effectively dispose of the fecal matter that accumulated in such quantities . . . [which were] a source of offense and a constant menace to the health . . . is but too common . . . with large and small manufacturing companies."[13] In New Britain, Connecticut, the state board of health found several iron, brass, hardware, wool, and cotton manufacturing firms employing a total of 3,075 workers. These people's feces and urine, and the water that carried such waste, all drained into Piper's Brook. The board of health estimated that the brook received on a daily average about 2,500 pounds of feces and 3,500 gallons of urine.[14] In 1876, Chicopee, Massachusetts housed 18 cotton mills, 33 foundries, 26 woolen mills and dye works, 3 paper mills, 6 gas works, a bleaching works, a hat works, 3 tanneries, and 34 sawmills and gristmills, all together employing some 8,984 workers, who disposed of their wastes in privies over ditches that eventually emptied either into the Chicopee or the Connecticut River, creating what the Massachusetts State Board of Health called a "very offensive stench from factory privies."[15] Chicopee was not alone in using privies that emptied into brooks and streams; several other towns up and down the Chicopee River system had factories that emptied sewage into the nearest stream. In the small manufacturing village of West Warren, for example, "nearly all the water-closets of the factories and mills, employing 2,340 hands . . . empty directly into the [Chicopee] river."[16] Clearly, nature's economy was out of balance.

Like the mills, the boardinghouses and tenements in the growing factory towns demanded some means of disposing of wastes other than the traditional privy hole or privy built over a ditch to the local stream. When the local streams into which this sewage flowed became too "offensive," the mills and mill towns connected their privies to sewer lines that directed the sewage into the larger rivers and streams. They hoped this method would dilute the sewage and flush it away.[17] As the Massachusetts Board of Health noted, "When people are massed together . . . sewers become a necessity."[18] In 1852, the Massachusetts Supreme Judicial Court noted that this was a new problem that needed to be addressed. "Prior to 1834 this subject was unsettled."[19]

Despite the claims in 1853 of Holyoke's Hadley Falls Company that "the land, rising back from the river, makes the sites of homes sufficiently elevated to secure them good air" and a "perfect system of drainage into the river . . . thus securing a great preventative of disease," the "brick sewers of one and two feet diameter" that drained sewage into the canals neither met the needs of public health nor "secure[d] . . . good air."[20] Smells, particularly in the summer, troubled the city of Holyoke, and the two new sewers built in 1868 that dumped directly into the river or into a large open cesspool did not relieve the problem. Indeed, the open cesspool, which the city used for another six years, posed a significant health risk until a typhoid epidemic in 1874 spurred the company to drain it.[21]

Sewage problems were not confined to Holyoke but plagued all towns and cities, particularly the inland towns.[22] As populations grew, the Massachusetts Board of Health complained that "the temptation to cast into the moving waters every form of portable refuse and filth to be borne out of sight is too great to be resisted."[23]

The town of Keene, New Hampshire, on the Connecticut River tributary Ashuelot River, dumped sewage into a brook that ran through the it. What had been a clear mountain stream became by the second half of the nineteenth century tellingly known as "Town Brook Sewer." Eventually, it was covered and became a drain handling the town's rain runoff as well as its sewage. This eventually flowed into the Connecticut.[24]

The industrial towns of Holyoke, Chicopee, Springfield, Hartford, and New Britain daily dumped approximately 42.25 tons of fecal matter and 45,900 gallons of urine into the Connecticut River.[25] In its investigation of the tributaries that New Britain dumped sewage into, the Connecticut State Board of Health found Piper's Brook, "formerly a bright and beautiful rivulet abounding in trout and other fish . . . now little better than an open sewer."[26] It found that Gully Brook was "an elongated cesspool," and Park River "little better than a large uncovered cesspool."[27] Investigating Hop Brook, which ran into the Hockanum River and then flowed into the Connecticut just below Hartford, the board of health found that "the chief source of pollution to Hop Brook is the excreta from the mill employees. Nearly half a ton of fecal matter and 500 gallons of urine enter into the stream daily from the village and mill employees [some 1,800 people]."[28]

Industry and Industrial Waste

Human wastes were not the only problem plaguing the new industrial environment. Although in his 1887 speech, James Olcott expressed concern about cleaning up "sewage" from "families," it was waste from "factories"—the "raising of a polluted stream upon any body at the will of ignorant or reckless capitalists"—that outraged him the most.[29] Industrial wastes concerned Olcott because the mills generated a massive amount of them. Wastes from mills also angered Olcott because they violated what he felt was an earlier innocent relationship to nature for the profit of the few at the expense of the many. Olcott did not see industrial wastes as the byproduct of prosperity and progress, as Lyman did. For Olcott and the farmers he addressed in 1887, these industrial wastes were the castoffs of the privileged industrialists who profited and became wealthy while streams were polluted and farmers' animals became sick drinking fouled water. Olcott saw fouled water as a product of human action, not as an unintended consequence of the invisible hand of progress.

The industries that grew up in New England's river valleys needed water to power their machines. In some industries, such as papermaking, water entered into the industrial process itself as a raw material. Woolen and silk production required water to wash and process the wool or silk cocoons. Metallurgy industries needed water to wash and clean tools and equipment. Mills dumped these wastes into the nearest water system. As a result, many forms of industrial pollution plagued the Connecticut River Valley's waterways, as they did the other river systems of the region.[30] These wastes in turn affected the quality of the water.

The Massachusetts State Board of Health noted in 1872 that "the brooks and rivers offer naturally a convenient means of disposing of such waste matters as will float down stream, and of such as may be discharged, dissolved, or suspended in water. . . . Manufactories are located on riverbanks, particularly for the sake of the water-power, particularly on account of the desire to use the water in various manufacturing operations, and particularly because the running stream affords the opportunity of readily disposing of waste liquors and other refuse."[31] By the 1870s, waterways had added to their traditional functions of supplying drinking water, irrigating fields, providing transportation arteries, and powering machinery the role of carrying away wastes.

Nineteenth-century "anti–stream pollution" reformers believed that papermaking was the hardest on the region's water systems. Papermaking required two processes. First, the pulp, whether wood or rag, had to be prepared. The pulp was then spread into sheets through a process of rolling, drying, and pressing. Creating the pulp generated the greatest waste and the largest amount of water pollution. Before the 1880s, rags were the principal ingredient in making pulp, or "stuff"; later, wood pulp made up the raw materials in paper. To begin the process, all the materials had organic matter dissolved by application of an alkaline solution of lye–sodium hydroxide followed by heat and beating. The matters separated from the wood included resin, gum, fatty and coloring substances, and silica. Dirt, grease, and dye were removed from rags. To get rid of unwanted organic material, the raw rags were boiled in caustic lime (CAOH). Colored rags required a second soda boiling. Wood pulp was boiled in caustic soda and quick lime (NAOH). The waste from this process, consisting of dirt, grease, and oils from rags and organics, as well as caustic soda or lime, was then dumped into the river. During the boiling, the soda combined with silica and resinous matters to form a dark brown, almost black soapy water called "boils." To prepare 100 pounds of paper stock from white rags required 3 pounds of soda ash and 8 pounds of caustic lime. Colored rags needed 2 pounds of soda ash and 13 pounds of caustic lime.[32] Following the boiling, the material was cooled in a second wash of water and washed after the boils had been removed from the vats. Dumping the "boils" and then the "coolings" into a nearby stream often left it with a "froth or scum over its whole surface for long distances, and which has accumulated . . . to a thickness of four or five feet."[33] Once boiled and cooled, the material was washed and pulped, or turned into stuff. In this process, the fiber was subjected to the action of running water in rollers and drums. Liquid waste from the washing machines contained a considerable amount of suspended fiber. The material was then mixed with alum or sulfuric acid and bleached in chloride of lime. The bleached material was then mixed with water and small amounts of alum, soda, resin, and clay. The wastes from these processes all found their way into the waters near the mill.[34]

Although rag paper generated large amounts of toxic wastes, the volume of "impurities" generated in wood-pulp papermaking proved significantly higher.[35] While rags required only 2 to 3 pounds of soda ash to generate 100 pounds of paper stock, wood pulp required 100 pounds. Wood pulp used 60 to 100 pounds of caustic lime, while rags required 8 to 13 pounds.[36] Wood pulp grew in popularity to become the chief source material for paper in the second half of the nineteenth century. Paper mills in the northern valley, particularly at Upper White River Falls and Bellows Falls, in the 1870s began to switch to wood pulp for newspaper stock. The mills in Holyoke continued to concentrate on rag paper into the 1880s. In 1880, Holyoke built its first large wood-pulp paper company, the Chemical Paper Company.[37]

Paper mills, Holyoke, Massachusetts.

The switch from rag paper to wood-pulp paper not only increased the pollution dumped into the region's rivers and streams but also took its toll on New England's forests. At Bellows Falls and White River Junction, the International Paper Company opened wood-pulp paper factories that consumed hundreds of thousands of feet of wood for pulp paper.[38] Edward Everett Hale, in a romantic mood, at the end of the century noted that although New Englanders had been cutting trees since colonial times, lumber for construction demanded the largest and most formed trees, thus allowing the forests to grow back, while the "business of paper pulp [was] stripping [New England] of her magnificent forests." "Then alas!" Hale lamented, "Satan came walking up and down and he devised methods of making paper from wood pulp. . . . What follows is that you enter your forest with your axes in summer as you once did in winter, and you cut down virtually everything."[39]

Much as it deserved its reputation as the region's greatest polluter, papermaking was not the lone sinner. Wool processing had the next worst reputation for fouling New England's waters.[40] As a western Massachusetts doctor reported, "Woollen-mills are most polluting of any other than a paper-mill."[41]

Before raw wool fleece could be turned into usable yarn, lanolin oils in the fleece had to be removed with soaps and the wool bleached white with chlorine. After the wool had been cleaned and dyed, oils were reapplied and washed out again. This gave the wool the necessary softness and flexibility for spinning and weaving. The wool cloth manufacturing process used water ten different times. Raw wool was first scoured and washed in water and urine (in the amount of half the weight of the wool), soda ash, phosphate of soda, and ammonia salts. Pigs' dung was sometimes substituted for or combined with the urine. These were washed out of the wool and "dumped into the rivers or streams."[42] The wool was dyed in vats of dyewood, indigo, fustic camwood, madder, copperas, argol, alum, blue vitriol, tin crystals, and potash.[43] After this initial dying, the vats were emptied. "It is this waste material that discolors the streams so much and which causes the chief complaints by the inhabitants along the streams."[44] Wool workers then rewashed the wool and took it to a second dying vat with 260 pounds of dyewood and nitro-muriate (nitric acid or hydrochloric acid in some form), in which they boiled the wool, and then they emptied this second vat. To clean and dye one thousand pounds of raw wool required six thousand gallons of water. The wool was then carded and spun. Before weaving, it passed through a weak glue. The woven cloth was next washed in a mixture of urine, pigs' blood, pigs' dung, and soda. It was then fulled in a soap wash and fullers earth, which in turn had to be washed out. The water used to wash the dirt, lanolin, and various dyes, washes, and chemicals out of the wool went back into the rivers, laden with organic waste—oils—and chemicals.[45]

Although the paper and woolen industries received the greatest scrutiny from those, like Olcott, concerned about pollution, the cotton mills, especially the bleaching and dyeing processes, also generated industrial wastes. In bleaching, cottons mills dumped into the nearest stream alkaline, soapy liquids, solution of chloride of calcium, sulfate of lime, and traces of chloride of lime. Dyeing involved large amounts of organic material and heavy metals. In the early part of the century, cow dung or phosphate of soda and phosphate and sulfate of lime was used for preparing the cloth to accept the dye. In the second half of the century, manufacturers switched to soaking the cotton in heavy metals in order to make the organic dyes fast to the materials. In the 1850s and 1860s, textile mills used arsenate of soda as a preparatory bath for the cloth. By the 1870s, a mixture of cow dung and arsenate of soda, made up of arsenic acid and soda containing about 33 percent metallic arsenic, was used. This preparation was effective and economical, but also "a virulent poison."[46]

Despite their weight, many dyeing materials contained little actual coloring. A ton of madder was needed to create two and a quarter pounds of actual dye. In the dyeing process, "nearly the whole of these dyestuffs is refuse matter, which, partly in solution, and partly in the solid condition, [wa]s carried by the goit [the sluice-way] of the mill into the adjacent stream."[47] In the process of dyeing and making garancine (prepared madder), 25 percent of the used madder went into the stream as suspended waste, while 75 percent, rendered soluble, entered the stream in solution. As a result, the wastewater contained a large proportion of organic carbon and organic nitrogen in solution. The major dye pollutants generated in New England's cotton mills in the nineteenth century were madder, peachwood, logwood, sumac, cow dung, starch, and British gum. The major chemicals used were sulfuric acid, muriatic acid, soda ash, bleaching powder, lime, soap, and arsenate of soda.[48]

Over a single year, a medium-sized cotton textile mill in New England employing 250 workers consumed over 500,000 pounds of madder, 8,512 pounds of peachwood, 58,240 pounds of logwood, 17,696 pounds of sumac, 127,680 pounds of cow dung, 109,760 pounds of starch, and 42,560 pounds of British gum. It also used 280,000 pounds of sulfuric and muriatic acid, 112,000 pounds of soda ash, 31,360 pounds of bleaching powder, 67,200 pounds of lime, 98,560 pounds of soap, and 42,560 pounds of liquid arsenate of soda containing 833 pounds of metallic arsenic. Each year, this medium-sized mill flushed 600 million gallons of polluted water into the nearest river or stream.[49]

New England's silk manufacturers generated many of the same wastes cotton and woolen mills produced. In preparing raw silk for the dyes, the gum that adheres to the fibers was washed out with a heavy soap. It took twenty pounds of soap to prepare a hundred pounds of silk. Once clean of the gum, workers soaked the silk fibers in various compounds to make them accept the dyes, which were themselves mixed with a variety of toxic chemicals. "The waste, which is foul, [wa]s dumped."[50]

The leather industry supplied glove makers and boot and shoe companies but in the process contributed its unique pollutants to the region's waters. To prepare raw leather, nineteenth-century workers soaked it in vats of tannic acid made from the bark of sumac trees. Wastes from tanneries contained not only tannic acid but also organic materials and salts from the hides. In 1881, the tanneries of two large glove companies in Littleton, New Hampshire, dumped two hundred tons of waste, hair, animal tissue, and tannic acid into the Ammonoosuc River. The wastes spoiled ice in a local pond, and when the river flooded in the spring, the surrounding fields had "an almost incomprehensible amount of debris scattered over the entire surface. . . . In places this refuse material was evenly spread over the ground to the depth of an inch."[51] Local farmers complained the wastes rendered their most productive river-valley fields practically useless.

The metal industry developed along the length of the Connecticut River: in the northern valley around Franconia, New Hampshire, and Windsor and Springfield, Vermont; in the central valley around Chicopee and at the armory at Springfield; and in the lower valley at the armory works in Hartford and the brass and hardware works in New Britain. These factories demanded water not only to power bellows and hammers and to turn lathes but also to wash and cool their works. After the introduction of steam power in the second half of the nineteenth century, the metal industry increasingly moved away from water for power but continued to use it for cooling and cleaning.

Metalworking generated three major kinds of pollutants that found their way into the water system: burnt particulates (cinders, scoriae, and furnace ashes), acids, and metallic salts. Mill workers lubricated the bearings of the rolls in the rolling mills with tar or thick grease. The rolls were washed and cooled by water, which also picked up the lubricating tar and grease, as well as oxide from the iron. Ash and slag from coal used to heat the iron and run steam engines contained sulfur and arsenic, which poured into the air as particles in smoke, and into the water when disposed of as waste. The galvanizing of iron for wire or plate and the production of brass and brassware produced several "metalliferous liquors" as waste. Sheet iron, iron castings, and other iron products were cleaned by immersion in a bath of water and sulfuric acid. They were then sanded smooth and washed with water to ready them to accept a coat of zinc covered with a thick layer of sal ammoniac. Workers then washed surplus zinc and other chemicals from the sheets. In the process of pickling, iron was subjected to washes of sulfuric or other acids to remove oxide from the surface. Ironworks such as the ones in Franconia, where iron plows were produced, used acid to remove silicate crust or scales that adhered to the plows. Along with the jobs and economic prosperity, the metalworking industry also brought the region sulfuric acid, chlorine, and various suspended metals and filings, which were all dumped into the area's streams. As the Massachusetts State Board of Health noted in 1876, "It is understood, however, to be the general practice in these works to discharge the waste contents of the acid-baths [acids and sulfate of iron] suddenly into rivers or sewers, as the case may be, rendering the water of the former unfit for the support of fish-life, if not absolutely injurious to the health of man or air-breathing animals."[52]

Wastewater from brassmaking and brassworking contained salts of copper and zinc dissolved in acid water. In 1888, the Connecticut State Board of Health reported that the dumping of those wastes had "rendered the water of the streams wholly unfit for fish, the chief waste, sulfate of copper, being the most poisonous of any substance known to this form of life."[53] Sulfuric acid was used to clean the scaly deposits formed from soldering of copper tubing. In making chandeliers, brass was pickled in nitric acid and arsenic. In the electroplating process, a coat of silver, gold, or nickel was placed on another metal or alloy, while the original metal was boiled in potash-lye, then pickled in nitric acid, in sulfuric acid, and then sanded. The wastes from these processes were in turn dumped into the local stream or river. In New Britain in 1888, a series of iron-, brass-, and hardware manufacturers, as well as a wool and cotton mill, together employing over three thousand workers, poured into Piper's Brook and the Park River 420,000 pounds of acids, 20,000 pounds of lime, 95,000 pounds of alkalis, 9,000 pounds of alkaline salts, 7,000 pounds of metal salts, 300 pounds of alum, 2,500 pounds of cream of tarter, 25,000 pounds of soap, 50 pounds of aniline color, and 2,000 pounds of logwood, for a total of 580,850 pounds of waste daily.[54] The Stanley Iron Works in New Britain used sulfuric acid in its pickling vats to clean the iron, while cream of tarter and caustic soda were used in whitening and cleaning. Several times during the day, these vats of chemicals emptied into the local stream.[55]

Water pollution was not a concern only for farmers and citizens. It also concerned manufacturers. Local industries along the Hockanum River worried that "any addition to the already great pollution of the water" would endanger their businesses.[56] The Adams paper mills, which contributed a significant amount of pollution of their own, complained that they had located in Manchester because of the pure river water of the Hockanum and now had to spend $2,000 a year to clean the water.[57] And a woolen mill, one of the worst polluters of New England's waters, closed its doors downstream from New Britain "because the water was wholly unfit for cleansing wool."[58] In Lisbon, New Hampshire, meanwhile, the turbine waterwheel of a local mill was shut down by industrial wastes from glove manufacturers upriver, and a local paper mill had to "strain all of the water used in their various processes."[59]

Manufacturers emptied their wastes into the nearest stream because that was the easiest and cheapest means of disposal. In doing so, they externalized part of the environmental costs of their production process. Although doing so was the general practice of the time, nineteenth-century observers were well aware that pouring industrial wastes into the region's waters fouled them. As early as 1865, Theodore Lyman, as state fish commissioner, noticed that the race water leaving the mills was "all dirty." The water was so thick with pollutants "that fish could not get past."[60] Increasingly, New Englanders not only noticed pollution, they began to fight against it.

Pollution reformers like James Olcott were concerned about the pollution of the region's water because they believed that dirty water led to poor health and destroyed the aesthetics of their communities and threatened the livelihood of farmers. Water pollution was seen as an important public issue because, as the Massachusetts State Board of Health warned, "[It] reach[ed] to the very foundations of national health and prosperity." Pollution seemed to "press with great force upon the people," as the board of health noted, because pollution was still a relatively new experience for these New Englanders. It was a result of a change of which they were aware. "Some of our brooks which were but recently pure and undefiled are now polluted so that neither man nor beast will freely drink of them; and this change is insidiously taking place from year to year."[61]

These reformers were empiricists. They believed that science clearly established that polluted water created miasma, which caused of disease. As the Connecticut State Board of Health noted in its first report, "Year by year our knowledge increases of the amount of sickness and disease caused in some way by impure water, whether through its drinking water or through the effect of exhalation of polluted air we breathe."[62] It was not only scientists who were concerned. From the farmers along the Park River downstream from New Britain who complained that polluted waters and bad air rising from it made them and their cattle sick, to the statisticians and doctors who saw clusters of sick people in areas thick with smoke pollution and foul-smelling filth, more and more New Englanders believed that pollution created poor health and a bad environment.[63] These people observed that where there was foul-smelling water, there was also an increase in ill health. These farmers, health officials, and reformers concerned themselves not only with sewage wastes but with industrial wastes as well. They considered both to be destructive to the health of the environment.

Statistical studies by Lemuel Shattuck of Boston and Sir Edwin Chadwick of London had shown the relationship between crowded dirty urban areas and high levels of disease. By the second half of the nineteenth century, both "practical experience" and the science of the day pointed to the link between environmental degradation and poor health. For midcentury reformers, the sign of a clean environment was clear, clean water abundant with life.[64] Deacon Whittlesely, who lived along Piper's Brook, noted "the former purity of the water, and the abundant supply of fish it contained. Now," Whittlesely went on to complain, "the hay cut from the low meadows smells of sewage and the milk of cows drinking the water was spoiled."[65] When they went to test to see if the streams were polluted, the members of the Connecticut State Board of Health smelled and tasted the water and looked to see if live minnows and fish were abundant.[66]

When pollution that flowed into the region's waters, in the words of the Massachusetts State Board of Health, "rendered the water of the former unfit for the support of fish-life," it was understood that it would also be "absolutely injurious to the health of man or air-breathing animals."[67] Reformers argued that rivers "wholly unfit for fish" were unhealthy and in need of public action. Industries that dumped pollution, like the woollen industry, "[seem] to kill the fish, thereby causing much nastiness."[68]

New England's Lost Innocence

By 1860, the cutting of the region's forests and damming of its rivers and streams had compromised New England's waterways. The world of mills and trains, markets and hustle, soot and obstructions that Henry David Thoreau dreaded was coming into place in 1839 had firmly settled over the region. A quarter of a century later, New England's rivers and streams were compromised not only by mud and freshets' wash but by sewage and refuse. A study of the Connecticut River found that the "river receives large amounts of sewage from Holyoke, Chicopee, Northampton, Springfield, Hartford, Middletown [cities with an aggregate population of 170,000 by the late 1880s], and many smaller riparian towns, sewer directly into it."[69] The river whose "purity, salubrity, and sweetness of its waters" encouraged Timothy Dwight to call it in 1803 "the most beautiful river," and his relative Theodore Dwight to dignify it with the title "the King of New England streams," had become a river of concern, "unfit for drinking at any point."

James Olcott reminded his agricultural society listeners that the region had not always been polluted. "Our streams were . . . clean . . . when our ancestors found them two or three hundred years ago." But Olcott also warned his audience that much needed to be done to save those streams from "soiling." "We want to make it impossible that any man totally ignorant of the insane conditions which have changed our New England climate from one of the most salubrious to one of the worst in the world, can be sent to our legislature, or to be entrusted with any official position in the state." Olcott realized that powerful interests were opposed to "anti-stream pollution," but he also poetically reminded people that although the development capital for the region's mills came from Boston, the men who helped pollute the area's waterways (including Holyoke Water Power Company president Alfred Smith) "were but yesterday barefooted boys, fishing these same brooks in the most enterprising manner."[70]

Olcott wanted the farmers of Connecticut to move forward against pollution to get rid of "this relic of recent industrial barbarism without the pale of civilization."[71] To agitate against pollution was one thing. To deal with the problem was another. Olcott saw pollution as something that affected not just the urban environment. He reminded his audience of farmers that the environment of the city also reached into the country, and that the struggle for clean water was one that would "unite our best people in city and country." He also believed that ending pollution would require public action by the state to bring "lasting welfare [to] the whole commonwealth."[72]

The problem of pollution faced many New Englanders. It also brought forth a variety of answers. For New England antipollution reformers, the answers emerged out of the metaphor of "the economy of nature" suggested by the New Hampshire Board of Health. If the economy of nature and its invisible hand were "swept away" by man's "artificial constructions," then the obvious solution to the problem was the more visible hand of the state so pilloried by Adam Smith and his followers. Smith found in the self-regulating working of supply and demand an economy that functioned naturally and best without the agency of the state. Arguing against such a program, these antipollution reformers of the second half of the nineteenth century saw a new role for the state in a world where nature's economy no longer functioned smoothly and indeed needed the more visible hand of the state. But how aggressive would that role be, and what could the state do about the problem? These were the questions the next generation of New Englanders would face. In grappling with these questions, New Englanders spearheaded what became known in later years as progressive conservationism.[73]

4

Pre-1860 Responses to Change

Views of the Public Good

Enthusiasts for Progress

Timothy and Theodore Dwight saw the coming of the mills and manufacturing as an example of industry and energy among the people of New England.[1] The Dwights looked at the development of industrialization in New England at its early stages. For them, mills and manufacturing signified increased wealth and employment, a belief shared by many New Englanders. Theodore Lyman III believed that without manufacturing, New England would be poor, miserly, and ignorant. Not all New Englanders were as optimistic about manufacturing, but those who were had the support of the courts, and significant influence in the highest offices of the region.

Nineteenth-century New Englanders of all stripes realized that a rural agrarian society was giving way to an urban industrial society. They understood that this transformation not only affected the immediate environment of cities and towns but also reached into the surrounding countryside, to the farms along the river valley, up to the forests of the hills and mountains, and into the waters of the rivers, brooks, and streams that flowed away from the factories, towns, and cities.

The Courts, Mills, and the Public Good

Dams dotted the late eighteenth-century countryside. But the dams, even the small eighteenth-century ones, also flooded fields and blocked the migrating fish.[2] In the eighteenth century, farmers and fishers whose fields were flooded by the mill dams or whose fishnets were empty because of a dam blocking the migration of anadromous fish often took direct action against the dams. The judges of the Massachusetts Supreme Judicial Court noted that if a dam was seen as a common nuisance, "any individual of their private authority might tear it down at any season."[3] In 1799, Elijah Boardman and several of his Connecticut River Valley friends climbed onto Joseph Ruggles's mill dam and ripped out the upper portion, which had raised the dam an additional ten inches and flooded fifty acres of land. Boardman admitted to destroying Ruggles's dam but claimed the right to do so on the grounds that the dam was a public nuisance.[4] In 1827, Oliver Moseley and twelve of his friends entered Horace White's mill dam site and tore down the dam across the Agawam River, claiming that the dam was a nuisance. The early court dockets were full of these cases.[5] Common law historically protected property owners from having their lands flooded or access to fish denied. It was in this spirit that Thoreau, in thinking about the blockage of migratory fish by the dams at Lowell, wondered what a "crow-bar [would do] against the Billerica dam."[6] But Thoreau was a generation too late to take his crowbar out against the rocks of the Billerica dam.

Mill owners were already working to keep the Thoreaus of the world from wrecking dams. They argued that the public good of their mills was of greater importance to the community than a few acres of flooded fields or a few migratory fish. If every angry farmer took a crowbar to a dam or brought a lawsuit every time he felt aggrieved, mill owners claimed, they would never be able to run their mills. In response to the mill owners' pleas, the Massachusetts legislature passed the Mill Acts in 1795 (1796, ch. 74).[7] These acts effectively allowed the mill owners to become tenants of the flooded land. The mill owners would pay the landowners.[8] The passage of the Mill Acts did not stop citizens from attempting to exercise what they believed were their common-law rights. When they did, they oftentimes ended in court. The meanings of those actions and public rights were thus ultimately worked out in the courts. In making rulings on these cases, judges not only adjudicated particular conflicts but also contextualized the law. Their interpretations of common law also created a basis for an understanding of property and property rights that fit the newly emerging industrial society.[9]

When citizens took common-law action or went to court claiming common-law grievance against the mills, the judges of the supreme judicial court tended to interpret the law in such a fashion as to favor the mill owners, as can be seen in the 1805 ruling from Justice Theodore Sedgwick that the benefits of mills limited common-law rights.[10] By 1814, Sedgwick's concern about the public good of mills became entrenched in law. In a case brought by Abel Stowell against Samuel Flagg, Isaac Parker, the chief justice of the Supreme Judicial Court of Massachusetts, ruled that the legislature passed the Mill Acts "with a view to favor the owners of mills."[11]

Mill dams not only flooded meadows and fields but also obstructed the migration of fish that local farmers and commercial fishers depended on for food and income. Blocking that migration was a "great injury of the public, an evil example of all others in like cases offending against the peace and dignity of the Commonwealth."[12] Justice Theophilus Parsons, writing in 1808, felt that history and common law spoke on that subject. Building a mill and dam, he argued, implied limitations. "One was to protect private rights, by compelling [dam builders] to make compensation to the owners of land above for any damages occasioned by overflowing their lands, another was to protect the rights of the public to the fishery. Therefore every owner of a water-mill or dam holds it on the condition, or perhaps under the limitation that a sufficient and reasonable passage way shall be allowed for the fish. This limitation being for the benefit of the public."[13]

Although the Massachusetts Supreme Judicial Court voiced the idea of protecting fish migration, the judges were also disturbed by the dangerous precedent of citizens with crowbars tearing down mill dams. In its ruling of 1807, the court noted that mills obstructing fish passage should be regulated by legislative acts rather than by common law. If the legislature passed statutes regulating mill dams and fishways, then conflicts would be handled by adjudication rather than by individual acts.[14]

In 1818, the legislature passed legislation that theoretically would protect the migratory fish from obstructions to their travels, yet certain mill dams were specifically excepted.[15] And as Enoch Chapin of Hadley Falls found to his pleasure, the justices on the supreme judicial court did not believe the migration of fish to be of greater importance than mill dams. In 1824, when Enoch Chapin attempted to rebuild a dam across the Connecticut, some "gentlemen interested in the fishery of the Connecticut River" met at Warner's Coffee House to oppose the new dam.[16] Chapin built his dam anyway and was indicted for preventing the passage upstream of fish. The local jury found Chapin guilty of a nuisance under common law. Chief Justice Isaac Parker overruled the lower court, arguing that the dam could not be ruled a nuisance at common law because the long history of legislative acts had essentially altered common law. "Though a source of profit and support to multitudes of people [fish were] found also to be a fruitful source of disorder and contention growing out of the conflicting rights." Thus "regulations . . . deemed . . . expedient, qualify and restrain private rights on these subjects . . . [and] exclude the common law entirely in relation to them."[17] Although Chapin won at the higher court, he altered his dam to allow a fishway in the hope of pacifying those who fished in the Connecticut.[18] By looking to the fish acts as the means to limit fish obstruction, the courts limited the ability of individuals to get relief by claims of common-law rights, while at the same time the court upheld the principle that dams could not block fish migrations.[19]

Gristmills to Textile Mills: A Changing Conception of Public Good

Recognizing that mills and mill dams were necessities of life, and realizing that mill dams would flood fields and hinder fish migration, the legislature attempted to legislate requirements that mill dams remain open during fish migration season or provide fishways.[20] As Thoreau noted, "Those who at that time represented the interests of the fishermen and the fishes, remembering between what dates they were accustomed to take the grown shad, stipulated that the dams should be left open for that season."[21]

Although the original intent of the legislature was to favor mills for the immediate benefit of a predominately rural community, increasingly, judges interpreted the legislative acts as favoring mills themselves.[22] And the mills that they favored differed significantly from those built when the Mill Acts were passed. By the 1820s, the mill dams built along the Connecticut and other New England rivers ran spinning frames and weaving looms to manufacture cloth for sale in the cities of the East and as far away as the western frontier and the Deep South cotton plantations. Yet "the encouragement of mills [remained] . . . a favorite object with the legislature; though the reasons for it may have ceased the favor of the legislature continues," the court noted in 1827, much to the chagrin of Jacob Upham, who was found guilty of trespassing onto the property of the Wolcott Woollen Manufacturing to forcibly stop the flooding of his field.[23]

Lemuel Shaw and the New Order

As mills and mill dams were changing, judges' expanding notion of mill rights also reflected the growing role of judges as makers of law. These judges made law in a fashion that particularly favored the large mill owners. The leading figure in this movement was Judge Lemuel Shaw.[24] In 1830, Lemuel Shaw was appointed chief justice of the Massachusetts Supreme Judicial Court. Shaw, son of a pastor from Barnstable, Massachusetts, was born in 1781, educated at Harvard, and admitted to the Suffolk bar in 1804. In 1811, he was elected to the state legislature and for most of the next two decades served in either the state house or the senate. Shaw believed that the law should not impede economic development. In his decisions, he argued that not only did the legislature favor mills but the courts should judge the value of those mills for their contribution to the progress of the society. In 1825, when Silas Bemis attempted to use his mill privilege for a small gristmill, he had to contend with a large cotton manufacturing company whose mill dam flooded Bemis's mill privilege. Claiming to have the more ancient right, Bemis demanded that the new textile corporation dam be torn down as a nuisance. Shaw ruled that since the cotton textile company was a "very expensive manufactory," and "comparatively the dam site and mill privilege of [Bemis] [we]re of small value," Bemis would have to settle for damages and could not have the company's dam reduced.[25] In 1836, Chief Justice Shaw went even further to argue in *William Ashley v. Harlow Pease* that the courts not only should see the Mill Acts as exempting mill owners from common-law action, but they should view the acts in such a way as to allow for the changing economy. "The former construction [allowing the mill owners to use the waterpower as they see fit] will be more favored, because in general it is most beneficial to the grantee, by allowing a latitude of choice in the use he shall make of it . . . because such construction is most favorable to the general interests of the community by encouraging enterprise and promoting public improvements. It is better adapted to the growing and changing wants and the ever varying pursuits of an active community."[26]

When in 1842 Fisher Thayer converted his sawmill to textiles, he repaired an old dam and raised the water level to power the new machinery. Thayer also kept the dam closed all year. William Cowell objected and took Thayer to court. Cowell believed that the new dams were different from the old dams. Reflecting the older worldview, Cowell claimed that "according to custom of the country, a sawmill, or other mill has been kept up in the winter only, and the mill owner has uniformly been accustomed to drawing off the water sufficiently early in the spring to allow the growth of a crop of grass and to continue it down until the hay is cut and got in." Cowell complained that Thayer's use of the dam was significantly different, constituting "a new use not within the mill owner's prescription." But the supreme judicial court saw it otherwise. Chief Justice Shaw, arguing for the advantage of economic growth and seeing the interest of the state in industrial expansion, stated that "where one has acquired a right to raise and maintain a head of water by using it for one purpose, he may apply it to another, he may substitute a cotton factory for a sawmill and the like on the grounds that any other rule would put a stop to all improvements. So someone under the same logic may improve his claim, improve his machinery . . . and keep water at a constant height."[27] This was the world of the future, of cotton mills replacing sawmills and textile workers replacing farmers. It was also a world in which judges became arbiters of the public good.

As Chief Justice Shaw noted in 1851, "in consideration of the advantage to the public to be derived from the establishment and maintenance of mills, the owner of the land shall not have an action for their necessary consequential damage against the mill owner, to compel him to prostrate his dam; and thus destroy or reduce his head of water."[28]

In an 1856 Essex Company case in which angry farmers and fishers demanded the textile corporation build a functioning fishway or lower its dam, Judge Shaw argued that the Essex Company also engaged in "an activity public in [its] nature, and designed to promote the public benefit, [and] that it was quite competent for the legislature to exercise the power of eminent domain, by authorizing them to take private property when necessary.... These objects were to establish a great water power for manufacturing and mechanical purposes."[29] Shaw, clearly of the opinion that manufacturing was of greater importance than fisheries, reflected on which of the two public goods, fisheries or manufacturing, might be of the greater importance. "Whether that public good, expected from the fishery, consisted in affording an additional article of food to the people, or an employment for labor, or otherwise, the legislature might well compare this with the public advantage, in affording increased profitable labor and means of subsistence, and various benefits, from building up a large manufacturing town, and decide as the balance of public benefit should preponderate."[30]

In 1853, Otis Howard took the Proprietors of Locks and Canals on the Merrimack River to court, complaining that the proprietors' dam was raised by flashboards (boards which are placed across the dam to increase water level) and that this flooded additional land of Howard's. Howard claimed that under the Mill Acts, the jury had the power to determine the dam's height and the season the dam could be kept open. Howard believed that if the Mill Acts were for the local public good, the local jury should be able to determine how high and over what period of time the dam could be kept up. Chief Justice Shaw rejected this argument. Shaw believed in progress and the role of law to defend contracts and property. He had little sympathy for local control or for seeing public good in local terms. He argued that "there is no ground upon which such a position can be maintained. Parties are bound by their tenures and contracts, and have their vested rights; the statutes are designed to declare and secure them, and not to defeat them by actions of [lower] courts and juries."[31] Shaw believed that economic progress was in the interest of the state and that local communities could disrupt that progress by "exercising" the "rights of the public."[32] Shaw felt that the purpose and role of the courts was to protect progress against such localism.

For the justices of the Massachusetts Supreme Judicial Court, increasingly what was seen as the good of the community was not the good of the local farmers or fishers or even mill owners, but rather that of the large economic enterprises that not only built large dams and mills but also built towns and even cities and employed not a few workers, but hundreds. In 1835, in *The Palmer Company v. Isaac Ferrill*, the court argued that the 1795 c 74 statute did not distinguish between one species of mill and another, whether it was a "saw-mill which may employ a few laborers living at a distance, or the working of a cotton factory, which may employ a great many persons constantly and whose employment may naturally be presumed to draw around it dwelling houses, school houses, taverns, churches, banks, and other establishments ordinarily resulting from the increase of population."[33] Although the courts argued that the statute did not make a distinction because there were no large cotton factories when the 1795 statute was passed, by blurring the distinction in the nineteenth century, the courts were favoring the larger factories. This was a fact to which Silas Bemis could testify.

By the 1830s, the shifting understanding of what water was, of water rights and of waterpower, became even more abstract. The courts had to deal with a different world in the 1830s than that of gristmill operator Seth Spring, whose dam in 1804 flooded Sylvanus Lowell's field. There, the courts ruled that "the private rights of private property are never to be invaded but for obvious and important purposes of public utility. Such are all things necessary to the upholding of mills"; but the court also ruled that Spring couldn't flood Lowell's fields from the twentieth of May to the twentieth of November because it was not necessary.[34]

The court dockets were only reflective of a larger pattern of response to the change occurring in New England. People and legislatures responded to the environmental changes confronting them, individually and reactively. They challenged change once it had occurred, or they looked to some type of individual redress. But the world they challenged and the incidents they wanted redressed were changing faster than individual and reactive strategies could accommodate. The forces behind the changes were greater than they realized, and the changes were more widespread and far-reaching. The corporations building the dams and the mills were no longer local gristmills and sawmills, but rather large, heavily capitalized entities employing operatives. The courts (as Silas Bemis discovered) were reluctant to find for someone with only right on his side against a corporation with extensive factories that gave employment to hundreds. Increasingly, the courts saw public good and public interest as equivalent to corporate interest and corporate expansion.

The Naturalists and Progress:
Jerome Van Crownishield Smith

Court judges and industrialists were not the only ones articulating a view of individualism and progress. Jerome Van Crownishield Smith, one of the leading naturalists of the time, shared these beliefs. In his *Natural History of Fishes of Massachusetts, Embracing a Practical Essay on Angling*, he noted in 1833 that "in a country like ours, full of resources, . . . any individual with common skill and prudence, if he but throws himself upon the current will be sure of being borne on to prosperous results."[35]

Smith was born in 1800 in the rural community of Conway, New Hampshire. He graduated from Brown College in 1818. Smith moved to Boston upon completing his medical studies in 1825 and became a professor of anatomy and physiology and was appointed port physician, a position he held for the next twenty-four years.[36] In 1854, he become mayor of Boston. Smith wrote books on history, on his travels, and on popular science. Among his other publications were a book on beekeeping and several works on fish and fishing in New England.

Smith appreciated nature, but he also believed in progress and science. Smith's natural history of the commonwealth stressed the previous abundance of the state's waters. Yet he noted that "the remaining fisheries of the commonwealth . . . are in a great measure losing and in some instances [have] lost their importance. The beautiful salmon which Isaac Walton accounted the king of fish is a rare visitor to our waters."[37] Salmon, Smith noted, were particularly abundant on the Connecticut River, which "ha[d] been more distinguished for this fish, than any other river in Massachusetts, but yet they are becoming more and more scarce from year to year. Locks, steamboats, the common business of navigation and the above all increasing settlements, conspire to interrupt the progress of the salmon toward the head waters."[38] But Smith did not decry development and the depletion of the Connecticut River's fish populations. For Smith, the reduction in numbers of the region's fish was an unavoidable consequence of development and progress. The answer was not to have the state interfere in the affairs of men but rather for the workings of a sound political economy to adjust the needs of man and nature. If salmon lost their importance to Massachusetts in the 1830s, they were still in abundance in the less settled areas of Maine.[39] "Factories and sawmills have done their part towards the work of extermination. . . . But though much diminished from these causes, there are more or less waters all over this state where the fish live."[40] Smith believed that progress could not, nor should not, be halted. If the fish in the settled regions of New England were incompatible with progress and development, then fishers should look further afield to areas not yet settled. In settled areas, Smith assumed that "wise political economy" would encourage "gentlemen owning estates on which there are fine basins of water to . . . stock them with trout."[41]

Although Smith noted that it was unfortunate that manufacturing interests had caused the decline of fish, laws and statutes that restricted fishing or forced manufacturers to curb their activity were in his mind counterproductive, contrary to the natural workings of the economy and civilization, and ultimately futile. Rather than fight the manufacturers,

Smith suggested, the "patriotic . . . angler" should "direct his thoughts and steps to a more favored scene" where the fishing is still good.[42]

Smith believed in Adam Smith's invisible hand. He did not believe that humans could protect fish. "Protective laws are perfectly useless, unphilosophical, and at variance with the grand scheme of nature which provides for the necessities of all organized beings, and sustains the existence of their species, under all changes, incidents, and circumstances."[43] Smith felt "it would be utterly impossible to exterminate the species"; natural law would eventually balance out human activity with nature's fecundity. The loss in one place was complemented by a gain elsewhere. "Dams, breakwaters, etc. across rivers," Smith argued, "were the results of civilization, and the fishes may forsake these streams, but their loss is trifling, at any particular locality, when compared with the advantage arising from the improvements of their solitary haunts. As animals recede before the inroads of civil life, so do the fishes and no human law can restrain them."[44]

Jerome Smith was suspicious of state intervention. "Nothing can be more absurd than the whole course of legislation on this subject. Look at the statute book . . . and find as many unphilosophical and absurd restrictions, on man's natural propensity for angling . . . and manifestly at variance with the design of our creator."[45] Smith believed that God had fashioned the world for human use. To use that world to advance civilization through the construction of dams, mills, and cities was God's plan. Surely God did not design the extinction of useful species. Thus, if civilization led to the elimination of fish at one spot, they could still be found, by those deserving, somewhere else. Yet the fishers Smith was concerned about were not the farmers along the Connecticut who could no longer supplement their diet with fish, but the gentleman angler who could, to Smith's mind, always travel farther afield to practice his sport.

The Critics: Henry David Thoreau

In 1839, Henry David Thoreau and his brother launched their homemade dory down the Concord for their two-week trip on the Concord and Merrimack. On that trip, Thoreau commented on the loss of fish in the Concord and the flooding of farmers' fields due to the building of the dams at Lowell. If the new dams, factories, and mill towns were inevitable for Smith and represented progress and the public good for Shaw, just such a change distressed Henry David Thoreau. Unlike Smith, Thoreau had no faith in progress and no enthusiasm for the new industrial age. "I cannot believe that our factory system is the best mode by which men may get clothing."[46] Yet the factories kept being built; more "populations sufficient to form a village" kept clustering about them, and the dams kept rising higher and higher. Smith and Shaw saw the factories and each new machine as improvements, for they allowed for the production of more goods. Thoreau did not believe that more things were a public good.[47] "While civilization has been improving our houses, it has not equally improved the men who are to inhabit them."[48] In rejecting the idea that the production of more goods was in itself a public good, Thoreau also asked what were the costs of this new production system to the environment and to the way of life of traditional farmers and fishers. Thoreau asked who spoke for the fish.[49] Although he positioned himself as the spokesperson for the fish, he also lamented the conditions of the fishers. "One would like to know more of that race, now extinct, whose seines lie rotting in the garrets of their children, who openly professed the trade of fishermen, and even fed their townsmen creditably, not sulking through the meadows to a rainy afternoon sport."[50]

It was not just the fishers whom Thoreau saw harmed by the dams. "At length it would seem that the interests, not of the fishes only, but of the men of Wayland, of Sudbury, of Concord, demand the leveling of that dam. Innumerable acres of meadow are waiting to be made dry land."[51]

Thoreau's response to the corporations and factory system was the return to a simpler life. "Simplicity, simplicity, simplicity," he enjoined. Thoreau rejected society's growing obsession with ownership. "Enjoy the land," he suggested, " but own it not."[52] Yet faced with what seemed to him the inevitability of the dams and the pervasive ideology of progress, Thoreau had little to offer but a critique of that ideology and a hope that in a thousand years the dams would collapse and fish would once again run in the river.[53] Thoreau shared with the other romantic Transcendentalists the idea that nature was a reflection of a larger divine plan. Unlike Smith, the Transcendentalists were not sure that man's transformation of the natural led to a more perfect world. Thoreau was probably the Transcendentalists' most suspicious critic of progress and development.[54] Although perhaps America's most recognized and trenchant opponent of modernism, Thoreau was not the only New England intellectual distressed about the changes in the environment.

Lemuel Shattuck and the Health Reformers

Thoreau focused on modernization and consumption and the impact of factories, dams, and logging on the countryside. Other New Englanders were concerned about the urban environment and its impact on the region's citizens. Increasingly, doctors and a growing group of statisticians began to publicize their concern over the connection between the filthy conditions of the urban environment and rising death rates. They noted that where there were filth, stench, and dirty water there was also a high prevalence of disease, particularly typhoid, dysentery, cholera, and consumption. For them, it was common sense to believe that the relationship between filth and disease was causal.

Although anxiety about public health was intensifying, it was not new to mid-nineteenth-century Americans. Epidemics of smallpox, yellow fever, typhoid, and cholera had already alerted Americans to the dangers of collective living.[55] In 1840, the city of Boston asked Lemuel Shattuck, a bookseller, genealogist, and member of the city council, to compile an analysis of the city's vital statistics. Shattuck had already established his reputation as a statistician. In 1839, as a member of the Massachusetts Historical Society and the American Antiquarian Society, he helped found the American Statistical Association and served as its first secretary.[56] In his report on Boston's vital statistics, published in 1841, Shattuck pointed out that Boston's health, as reflected in mortality records, had been deteriorating since the 1820s.[57]

Shattuck shared the widely held "anticontagionist" theory of disease. Anticontagionists believed that bad air or miasma caused disease.[58] Shattuck was not alone in his concern about the state's sanitary conditions. Edward Jarvis, a doctor and statistician from Dorchester, also believed that poor sanitary conditions contributed to disease and that a survey of the state's sanitary conditions would reveal the need for action.

Jarvis had support in this from the American Medical Association. At the AMA's annual convention in Boston in 1849, Joseph Curtis, a physician from Lowell, reported on his survey of sanitary conditions in Boston and Lowell. Curtis found mortality rates in those cities significantly higher than the state average. Curtis also noted that although Lowell was smaller than Boston, it had even greater mortality rates. Curtis believed that not only was there a link between filth and disease but that manufacturing also contributed to ill health.[59] In 1849, he reported to the Massachusetts state legislature that poor sanitary conditions and the deteriorating physical environment were responsible for higher mortality rates.[60]

Pressure from the state medical association and the American Statistical Association, along with the agitation from Curtis, Jarvis, and Shattuck, led the Massachusetts legislature to authorize an investigation of the state's sanitary conditions. In 1849, Governor Briggs created a commission of Shattuck, now a member of the legislature from Boston; Jehiel Abbott from Westfield along the Connecticut River; and Nathaniel Banks from Waltham to do the survey. Shattuck was appointed chair, and although Abbott was particularly concerned about health conditions in the inland river communities such as Springfield and Chicopee, both Abbott and Banks deferred to Shattuck, who wrote the report for the committee.

In 1850, Lemuel Shattuck gathered together the mortality figures of the state in the *Report of the Sanitary Commission*. He interpreted the statistics that he compiled as showing the impact of the environment on disease. The health of the state required that the community clean up the sources of poisonous air.[61]

Two thousand copies of the report were circulated around the state, and Shattuck made sure that others went out to reformers in other states and abroad. For many doctors, Shattuck's report was a call to action. Disease was caused by poor environmental conditions, and that demanded state action. But despite the initial enthusiasm that greeted the report, especially among reformers, little concrete action followed its publication.

Shattuck's work did point to a link between environment and disease. It also traded upon long-held suspicions that undermined its effectiveness. Although Shattuck wanted his report to focus on the need for public action to clean up filth, its elitist and paternal tone also gave support to those who believed that filth was a product of weak moral character. He believed that because disease was a product of filth, it was in human hands to do something about disease, and thus the state needed to act to improve public health. Others, though, tended to believe that filth was the product of human ignorance and moral weakness and hence the responsibility of the individual and not the state. Nativists, particularly, were inclined to place the blame on the personal habits of immigrants, especially those coming from Ireland. The anti-immigrant activists looked to immigration restriction rather than public cleanup as the needed remedy. Battles between the nativists and the antinativists and later conflicts over slavery shunted to the side Shattuck's report and call for political action, for at least another decade and a half.

George Perkins Marsh: Naturalist

Although many nineteenth-century New Englanders were aware of the changes occurring in their environment, it was a Vermont lawyer, politician, and amateur naturalist who saw the changes as part of a package of forces tied to the culture of the society. Through their interactions, these forces were transforming the region's environment. In 1856, the same year that Massachusetts began a suit against the Essex Company for obstructing fish on the Merrimack, fishers and farmers along the Vermont side of the Connecticut River petitioned the Vermont legislature for something to be done about the absence of the traditional spring fish runs up the Connecticut. Although salmon runs had ended almost fifty years earlier, shad had continued up the Connecticut into Vermont until the building of the Holyoke dam in 1849. As Vermont's rural economy became increasingly more tenuous in the 1840s and 1850s, particularly for the more traditional hill farmers, the lack of spring fishing, a source of protein and long a welcome social relief from winter isolation, gave rise to protests from valley farmers. Vermont legislators heard these protests and appointed George Perkins Marsh state fish commissioner in 1857 and asked him to study the problem.

Like many of his generation, Marsh identified outdoor life with manhood and manly virtues. In a common refrain, Marsh lamented that the youth of New England had "less physical hardihood and endurance . . . our habits are those of less bodily activity. [T]he sports of the field, and the athletic games with which the village green formally rung . . . are now abandoned." In the highly gendered view of nineteenth-century society, Marsh was convinced that New Englanders' "more thoughtful and earnest" life made them "more effeminate," a condition that threatened "our rights and our liberties."[62]

If modern cultivated life was sapping the manhood out of New England's youth, hunting and fishing could restore it. "The chase is a healthful and invigorating recreation. . . . The courage and self-reliance, the half military spirit . . . which it infuses are important elements of prosperity and strength in the bodily and mental constitution of a people."[63] But, alas, cultivated and civilized social life, the very forces that were undermining New England's manhood, were also destroying the potential for restorative outdoor sport. "The final extinction of the larger quadrupeds and birds as well as the diminution of fish, and other aquatic animals is everywhere a condition of advanced civilization and the increase and spread of a rural and industrial population."[64]

Many a proponent of romantic nineteenth-century manhood might have written as much. Indeed, Jerome Smith had made a similar point in the 1830s. Marsh, however, understood that New England's traditional game depended upon an ecological system that had been disrupted by human settlement. Marsh saw a system of interdependence that had been radically altered by the cutting of the forest, which raised "the annual mean temperature" of the streams, increased the violence of "spring and autumnal freshets" that "sweep down rural and their eggs, and fill the water with mud and other impurities," and deprived the system of the "abundance of . . . insects" for those fish and animals on the lower rungs of the food chain.[65] "Human improvements," Marsh noted, "have produced an almost total change in all the external conditions of piscatorial life, whether as respects reproduction, nutriment or causes of destruction."[66] Marsh was pessimistic about the possibility of restoring the region to anything approaching its original form. "We must, with respect to our land animals, be content to accept nature in the shorn and crippled condition to which human progress has reduced her."[67] Unlike Smith, who saw the losses of fish due to civilization as balanced out by gains elsewhere, Marsh saw man's handiwork as diminishing nature. Marsh had a view of civilization as deteriorating, which had "shorn and crippled" nature.

Both Henry David Thoreau and George Perkins Marsh had an acute sense of the natural world. Both were aware of and alarmed about the changes occurring about them and the loss of the traditional New England habitat. Like Thoreau, Marsh believed that modern industry contributed much to the destruction of the region's fish. "It is believed, moreover, and doubtless with good reason, that the erection of sawmills, factories, and other industrial establishments on all our considerable streams, has tended to destroy or drive away fish partly by the obstruction which dams present to their migration, and partly by filling the water with saw dust, vegetable and mineral coloring matter from factories, and other refuse which renders it less suitable as a habitation for aquatic life."[68]

Marsh differed from Thoreau in that Marsh understood that modernization in the form of the dam, canal, and factories alone was not responsible for the destruction of the fish.[69] While Thoreau vented his anger against the instruments of modernization, the factories and the railroads, Marsh understood that even such small changes as clearing a field for planting affected the conditions of life for the fish and wildlife of the area. Thoreau centered his critique of man on a critique of modernization. For Marsh, it was not so much modernization as human settlement itself, "the general physical changes produced by the clearing and cultivation of the soil," that spelled destruction of traditional flora and fauna.[70]

Although (like Thoreau) Marsh was pessimistic about the return of the region's animal diversity, his worldview was not antimodernist.[71] Marsh believed that "we may still do something to recover, at least a share of the abundance which, in a more primitive state the watery kingdom afforded."[72] To create that "something to recover, at least a share of the abundance," Marsh looked to the very agent, the state, for which Thoreau had such contempt. Conservation, Marsh believed, needed "to be promoted rather than discouraged by public even legislative patronage."[73] Marsh was impatient with the antistatist position of many of his generation. "The apothegm, 'the world is governed too much' . . . has done much mischief whenever it has been too unconditionally accepted as a political axiom. The popular apprehension of being over-governed . . . has had much to do with the general abandonment of certain governmental duties by the ruling powers of most modern states."[74]

Although Marsh believed that government action was an appropriate vehicle for restoring some of the original richness of the region's waters, he also believed that the culture and values of the region vastly limited the potential of the state for doing what needed to be done. "The habits of our people are so adverse to the restraints of game-laws . . . that any general legislation of this character would probably be found an inadequate safeguard."[75] Having little faith in the success of restricting hunting and fishing, Marsh looked to the science of pisciculture to restore fish to the region's waters. But even here, he had limited expectations. He believed that because the anadromous fish swam up the Connecticut through water within the boundaries of Connecticut and Massachusetts before they reached Vermont, until those states developed programs of experimentation and conservation, there was little the Vermont legislature could do other than encourage private fish-breeding experiments.[76] Marsh recognized the same major problem Thoreau had, perhaps more forcefully. But to both writers, it remained effectively intractable.

By midcentury, fishers in taverns mourned the loss of what they remembered as the good old days. Fewer fish meant fewer sources of food for the more marginal farmers. Farmers who remembered older times when game and particularly fish were more plentiful resented the loss of the earlier bounty. While New Englanders lamented the loss of turkeys, ducks, geese, and partridges, the loss of these resources was seen as beyond human control. Hunting and farming were understood to have caused the decline in wild game. For nineteenth-century New Englanders, hunting and farming were seen as natural activities, like eating and breathing—things that were inevitable and could not be controlled. The depletion of fish in the inland waters was seen as different. The decline of fish was not beyond human control. Humans intentionally built the dams and weirs. In the minds of many New Englanders, there was nothing inevitable about these dams. The older dams did not threaten the migration of fish. Fishways could be built around the dams. Particularly for New Englanders who viewed the new industrial world with suspicion (like Samuel Ely, "an old fashioned farmer . . . who didn't want to see the corporations control everything . . . [and] was sorry they came" and shot at the lawyer who tried to force him to sell his land to the Holyoke Water and Power Company), the destruction of the fish migrations by the dams was a wrong done to the land and the people.[77]

In the new world, dams blocked the migration of fish and flooded fields, waters flowed thick with human and industrial wastes, and little seemed to be done about it.[78] Henry David Thoreau suggested waiting a thousand years for the dams to fall down. George Perkins Marsh had a different solution. In 1857, it was a weak and whispered suggestion, but it was a suggestion that would increasingly come to the fore. For Marsh, private corporations that amassed power and wealth could only be countered by an agency of potentially equal significance, the state.[79] That agent was only in its incipient stage in the middle of the nineteenth century, but as the nineteenth-century natural world was transformed by ever more powerful and diffuse forces, more and more New Englanders looked to the state not as a reactive passive body, but as a proactive agent for the common good.[80]

5

Fish, the People, and Theodore Lyman

The Moderate Approach

On November 3, 1865, Theodore Lyman III handed his report for the River Fishery Commission to Massachusetts governor John Andrew. Then he headed north from Boston to Lawrence, where he met with newly elected New Hampshire governor Frederick Smyth and the fishery commissioners from other New England states. At that meeting, Governor Smyth, in Lyman's words, "undertook the high horse and said they would shut down the water from Lake Winnepiseogee [the nineteenth century name for Lake Winnipesaukee] if we did not give the fishways."[1] Smyth was no one to take lightly. As the son of a New Hampshire farmer, he knew the importance of fish to the rural diet, and as a founding member of the Republican Party, he was a politician of some significance.[2] Smyth was also under pressure from rural farmers in the Connecticut and Merrimack River Valleys who had depended upon spring fish runs and now faced depleted rivers. Regarding the New Hampshire governor, Lyman wrote in his diary: "The threats of New Hampshire were some of my business as commissioner."[3]

These threats were Lyman's business in more than just his role as fish commissioner. The waters of Lake Winnipesaukee fed into the Winnipesaukee River, one of the main sources of the Merrimack River, which provided the power for the mills at Lowell and Lawrence. Without that water, those mills could not function. Lyman enjoyed healthy returns on his holdings in those mills. He not only held stock in these companies and in mills in Holyoke, he was also on several of their boards of directors.[4] As he stated when he later ran for Congress, "I have been connected, and my father before me with the manufacturing interest."[5]

As a major stockholder, Lyman had reason to be concerned about the waterpower of the mills along the Merrimack. Yet when he met with the governor and fish commissioners, he thought of himself not as the representative of the manufacturing interests but as a scientist and public servant. It was a role for which he had been preparing for a long time.

Lyman came from an old Boston family. His grandfather, the first Theodore Lyman, was a Boston merchant who made his fortune in the China trade. As one of the Boston Associates, Lyman's father, Theodore Lyman II, had invested the money in textiles and focused his attention on public service.[6] From 1820 to 1825, he represented the people of Boston in the state legislature and served as mayor of the city in 1834 and 1835. As the son in such a distinguished family, Theodore Lyman III entered Harvard in 1851. He studied under Louis Agassiz there, along with Agassiz's son Alexander and Lyman's cousin Charles Eliot, who, with Lyman's support, would later become president of Harvard. After graduation, Lyman studied for an additional three years at the Lawrence Scientific School, focusing his attention on starfish. While studying starfish, he courted and married Elizabeth Russell, daughter of an East India merchant. In 1859, he became a resident fellow of the American Academy of Arts and Sciences, and he and Elizabeth went to Europe, where he furthered his scientific studies.[7]

Although the Civil War would later play a significant role in Lyman's life, in 1861 he was little concerned with the rumblings of secession and somewhat indifferent to the issue of slavery.[8] Lyman joined the Republican Party after the war but shunned its more radical elements.[9]

In the early summer of 1863, Lyman returned to Boston. He spent some time doing scientific work at the Museum of Comparative Zoology at Harvard and in the late summer wrote to General George Meade and offered his services as a volunteer on Meade's staff. Meade accepted, and within a month, Lyman was commissioned a lieutenant colonel by Governor Andrew. In the company of a trusted valet, he went off to the Union army.[10]

At the war's end, Lyman returned to Boston to manage his "farm," a Brookline estate on the edge of Boston's settled area. He picked up his scientific work and threw himself into public and charity service. Lyman also kept a close watch on his investments. He constantly visited the mills in which he had invested to look over new machinery, and he investigated property he had purchased or was interested in buying.[11]

Given his history, it is no surprise that Lyman believed in science and the scientific method, and he applied that method to his farm, as well as to his lab at Harvard. Scientific farming was a passion for many in Lyman's social class.[12] Indeed, his neighbors in Brookline—the Warrens, Coolidges, Danas, Lowells, Cabots, and Saltonstalls—had farms that they tended with all the care and precision of laboratory scientists.[13] Lyman brought a training and interest to this enterprise that surpassed even that of his neighbors. He was an active member of both the Massachusetts Society for the Promotion of Agriculture and the Massachusetts Horticultural Society. Lyman filled his diaries, notebooks, and scrapbooks with detailed information about the impacts of different fertilizers and seed mixtures on crop production, and with descriptions of the chemical effects of certain manures and additives to the soil. He experimented with combinations of additives to the soil and included a control mixture for comparison. The results (yields) were then carefully tabulated against the various mixtures.[14] These experiments reinforced Lyman's belief that science and technology, when carefully and rigorously applied, would lead to agricultural improvement. It was also a lesson that Lyman applied to life itself. He believed that discipline, application to detail, and science and technology would lead to a better life both personally and socially.[15]

Lyman did not have merely a gentleman's interest in science. He was a sincere scientist with impeccable credentials. At Harvard he studied under the tutelage of the nation's leading natural scientist, Louis Agassiz, and after graduation he continued his scientific studies both at home and then abroad.[16] Even when Lyman served under Meade during the war, he spent his days on leave indexing Harvard's Ophiuridae and Astrophytidae and working on a scientific paper on starfish reproduction. Returning from the war, he threw himself into his work at Agassiz's Museum of Comparative Zoology. Throughout his life, Lyman made almost daily trips to the museum to continue his studies, which led to several published manuscripts.[17]

Lyman's social set included not only the leading families of Boston but the city's leading scientists and intellectuals as well. Lyman regularly met in the homes of people like MIT's geologist William Rogers, or Harvard's Asa Gray or Louis Agassiz, or he socialized with naturalists Edward Wigglesworth, Hermann Hagen, and James Chadwick, historians Francis Parkman and Henry Adams, or author Oliver W. Holmes.

Like other Brahmins, Lyman believed in public service. He was a trustee of the Boston Farm School, a charity school, where he monitored the school's agricultural experiments, particularly the draining and cultivation of marsh land. Lyman also sat on the board of the Infants Asylum, to which he and his wife gave $20,000 at one sitting.[18] Lyman was a personal friend of landscape architect and Central Park designer Frederick Law Olmsted. When his Brookline neighbors grew concerned about providing a tranquil setting within the urban community, he joined them in "organiz[ing] a Park Commission."[19]

As it had for his father, public service for Lyman also involved politics. On September 24, 1882, Lyman's neighbors inquired if he would be interested in running as an independent candidate for the ninth congressional district. Lyman, along with other leading Boston figures, had been active in the campaign for civil service reform. Convinced that the state Republican Party would not endorse a strong civil service position, these reformers decided to strike out and run Lyman as an independent.[20]

Much to his surprise, the Democrats endorsed Lyman, which made his candidacy "serious."[21] As a result, he reported in his diary, he was "elected by 2,345 to Congress! Astonishment!"[22]

Lyman believed not only in civil service reform but also in "conducting the government business on business principles."[23] As a member of the Boston elite whose wealth came from the success of private industry, Lyman not only believed in the ideology of business enterprise but also thought that everyone shared this belief. "Americans are a business people," he wrote in his diary.[24] Despite support from the elite families in his district, who rallied around him, others in the ninth congressional district opposed him. They resented his paternalism and argued that he "allied himself with the representatives of corporate wealth."[25] By the end of his first term, it was clear that his reelection was in trouble. As he put it in his diary, "Lately everyone is against me!"[26] When the Democrats decided to run their own candidate, Lyman lost his reelection bid and returned to Boston.

Besides science and public service, Lyman had one other passion, outdoor sports, specifically hunting and fishing.[27] Nearly every autumn, he spent time duck and bird hunting outside Chicago.[28] When he was unable to travel west, he and a few friends traveled out of Boston to either the North Shore, the South Shore, or Cape Cod to hunt the less plentiful but more accessible Massachusetts avians.[29]

Lyman enjoyed hunting, but fishing was his greater passion.[30] He particularly favored fishing for sea trout or ocean fish (bass and blues) on the Red Brook in Wareham or on Cape Cod. From early spring through early summer, Lyman would take the morning train, once or twice a week to Cape Cod, fish most of the day, and return home at night.[31]

Although Lyman was a sportsman, he was not an Arcadian. His sports activities did not reflect a belief in a retreat into the wilds, nor did he advocate a wilderness preservationist approach.[32] He lived in a city environment and believed in urban life. He did not like the filth and oppressive conditions of the manufacturing towns, but he wanted his countryside within reach of the diversity and opportunity of the city.[33] Although he lived in suburban Brookline, he commuted daily to Cambridge or Boston, stopping for tea at any number of homes. His evenings were full of meetings, visits, social calls, plays, readings, or club events, all of which required the city. When Lyman did go out to the country to fish, he would take short daytrips. He was concerned about protecting greenery in the city and wanted to prevent the city from becoming "ugly," but his idea of prevention was to work toward urban parks.[34]

Lyman was an accomplished member of his social set. He could not help but be a representative of his social class, but he also represented science and what he liked to think of as modern, objective, and enlightened opinion. Faced with the problem of the region's declining fisheries, and particularly the ruckus being raised by New Hampshire and Vermont about the role of Massachusetts industry in that decline, Governor Andrew turned to Theodore Lyman III.

Fish and the People

The impressive eighteenth- and early-nineteenth-century record of legislative acts to protect fish not only reflected the importance of fish as food and as a commodity in the market place, it also reflected a world still very much linked to small traditional farms, gristmills, and sawmills that needed only a few head of water for only a few months out of the year. As the nineteenth century progressed, the progress came at the expense of the migration of fish. The increase of industrial activity for the region and the decline of traditional agriculture weakened interest in the local fisheries. Inland New Englanders focused more and more of their attention on manufacturing or producing agricultural goods for local markets. The role of fishing in this new world was not clear. The grandchildren of the farmers who Justin Alvord and Joseph Ely remembered coming over the hills every spring to catch fish found themselves working thirteen hours a day in the mills, which were powered by the dams built across those very rivers. The abundance of the fish runs remained a part of the popular memory, and the legislature continued to pass legislation protecting the fish, but in practice, the legislation was increasingly weakened and ignored.[35] The region's new canals and mills required waterpower year round and demanded a higher head of water and larger dams.[36] Textile manufacturers' mills and dams required heavier investments and demanded waterpower twelve months a year. The new massive dams built at the falls of the region's waterways not only transformed the countryside but also changed the nature of the waterways themselves. Contrary to the local dignitary's claim when the Barret Steamboat reached Bellows Falls, Vermont, that the Connecticut should be "dammed but never choked," for the fish that migrated up the region's rivers and streams, the big new dams did choke the river.[37]

For anadromous fish, these new dams meant an end to their yearly migrations, and for the fishers, it meant an end to a way of life.[38] When the dams were built at Turner Falls to provide navigation around the falls, the local paper noted that "malcontents . . . men who had netted salmon spring after spring in the deep pool . . . where the water was no longer churned by fin and tail . . . returned to the taverns to mourn the good old days along with those other unemployed, the fallsmen."[39]

Even before the big dams of Holyoke, Lowell, or Lawrence were built, migratory fish were already in trouble. Salmon need high levels of dissolved oxygen (DO) and clear pebbly bottoms to make their nests and cast their spawn. These conditions are found only at source streams of the major rivers.[40] By the beginning of the nineteenth century, with the construction of several dams across the tributary streams of the river, the salmon that were so prized in the eighteenth century no longer ran in the Connecticut or its tributaries. Although twenty- and thirty-pound salmon were caught at Bellows Falls in 1797, by 1820 they were gone.[41]

Shad are less demanding of their spawning spots. Shad continued to come upriver and to spawn, but their numbers fell as water was polluted.[42] Although the salmon runs had ended and the numbers of shad been reduced by midcentury, memory of earlier abundance, an abundance probably inflated by memory, remained among those who had fished the region's waters.

Anxiety about the region's fisheries did not emerge overnight. Those New Englanders who sat in taverns in towns like Hadley and Bellows Falls at midcentury remembered the prolific abundance of the region's waters and wondered what had happened to all those fish.[43] When Sylvester Judd interviewed local residents for his history of Hadley, and when the 1865 Massachusetts Legislative Joint Commission on Obstruction to the Passage of Fish in the Connecticut and Merrimack Rivers took testimony about the decline of fish on the Merrimack and Connecticut, the old-timers repeatedly commented on how rich the old fishing was before the building of the big dams. Old-timers tend to remember the past as richer and easier than the present, but their stories nonetheless gave credence to the claim that the building of the big dams had destroyed a resource that had since ancient times belonged to the people.

But that resource was increasingly compromised by the continued stripping of the forests along the northern watersheds of the region's rivers, the building of the dams, and the polluting of the rivers and streams. The industrial expansion of the first half of the century continued unabated in the postwar period, as more factories were built and ever more power was demanded from New England's waters. The companies established to provide the waterpower to the mills were anxiously trying to maximize all the potential power they could get from the falls of the rivers. Concern for fish and fishways, which might reduce waterpower, was not high on their agenda, even if they were a concern for fishers, farmers, and old-timers.

Yet the powerful new corporations nonetheless had to deal with the issues of fish and fishers' rights. In 1848, when the Essex Company built their huge thirty-two-foot-high dam at Lawrence, the charter required that the company build a fishway.[44] When the fishway failed to allow the passage of fish, the company successfully lobbied the legislature for a revision of its charter. The revision of the charter allowed the company to buy off holders of fishing rights damaged by the dam.[45] This did not satisfy all the fishers above the dam.[46] The anger of farmers and fishers against the Essex Company led the state legislature to pass a statute in 1856 repealing the charter revision and again spelling out the Essex Company's obligation to "make and maintain in or around their dam in Lawrence a suitable and sufficient fish-way for the usual and unobstructed passage of fish during the months of April, May, June, September, and October, in every year."[47] The Essex Company objected to this act, which it considered oppressive and unjust. The company argued that it had brought wealth and benefit to the state and should be seen as a public good and not penalized for a few fish. The company argued that the 1856 act was not so much an altering of its charter as a "tak[ing of] private property for public uses without compensation."[48] Despite the company's impressive attack on the 1856 act, petitions began to flood into Boston from farmers and fishers above the dam who opposed the company's push for the repeal of the act. The petitions were successful. The legislature stuck by its 1856 act, and the state sued the Essex Company for failure to maintain a "suitable and sufficient fish-way."[49]

The case came before Chief Justice Lemuel Shaw in 1859. The state attorney general, S. H. Phillips, argued that the revision of the Essex Company charter was "not a taking of private property for public use. It [was] rather a modification of the use allowed by law, or the restraining of a certain use which would be beneficial, for the public good."[50] Shaw agreed with the state that the Essex Company had clearly "failed to provide any new fishway." The question for the court to decide concerned the rights of the company to enjoy the property vested in it by its charter over and against the rights of the state. Shaw admitted that the protection of the fisheries was a public good. "The right to have these fish pass up rivers and streams to the head waters thereof is a public right and subject to regulation by the legislature." But Shaw was also concerned about other rights. He continued in his comments to note that in the exercise of one right, it was possible that another right might be transgressed. In such cases, Shaw argued, the legislative branch must make a decision about which right shall be of the greatest good.[51]

If there were any fishers in the court to hear Judge Shaw's statements to this point, they would have taken heart at such sentiments, for the legislature had fairly consistently defended the rights of fish migration. But their optimism would have been premature. Although Shaw noted that the 1856 act indicated that the legislature was not willing to allow manufacturing interests to supersede the existence of the fisheries, he took a different tack. He argued that in the years since the Essex Company's original fishway had failed, the legislature had in 1848 amended the company's charter to allow the payment of damages to those riparian owners of fishing rights above the dam. Having done so, the legislature had given the Essex Company "rights [which] have been acquired and become vested, [and] no amendment or alteration of the charter can take away the property or rights which have become vested under a legitimate exercise of the powers granted."[52] Thus the 1856 act was null and void."[53]

For those concerned about fish, Judge Shaw's 1859 decision in the *Commonwealth v. Essex Company* case was a defeat. But the decision did not put an end to the issue. In fact, concern for the fisheries increased in the late 1850s throughout the region. In 1856, the Massachusetts State Legislature authorized the appointment of a commission to investigate "the artificial propagation of fish as may tend to show the practicability and expediency of introducing the same into this Commonwealth under the protection of the law."[54] In 1857, the commission chair, R. A. Chapman, after consulting with Louis Agassiz, turned in his report to the governor of Massachusetts. At the same time, George Perkins Marsh delivered his famous "Report, Made under Authority of the Legislature of Vermont on the Artificial Propagation of Fish" to the Vermont legislature.[55] As noted, Marsh stressed that little could be done to restore the Connecticut River fisheries of Vermont if Massachusetts and Connecticut did not cooperate with "concurrent legislation"; Chapman himself pointed out that "this improvement in the fishery [in the Connecticut River] cannot be made without joint legislation on the subject in this state and in Connecticut."[56]

Science and Technology to the Rescue

The restoration of the region's fisheries confronted the legislature and the courts with the dilemma of balancing the political demands and legal rights of the manufacturers against the interests of the fishers and consumers of inland fish. It was a conflict rooted in limited resources. There were too few fish and too few water systems for all the competing demands for the resources. For many nineteenth-century intellectuals, science held out the solution to limited resources. Practical science for these New Englanders represented the answer to many of the problems that confronted them.[57] Gentlemen farmers like Theodore Lyman III believed that scientific agriculture would lead to improved production, less waste, and greater yields. For the agricultural modernists of the Connecticut River Valley, scientific agriculture was the means to save the region's farms.[58] If science could save New England's agriculture, perhaps it could save the region's fisheries too.

The ideology of improvement science and the restoration of nature's abundance was expressed by J. W. Patterson before the Game and Fish

League of New Hampshire in 1878. Patterson, following up on the ideas of Jerome Smith, admitted to his audience that "fish gradually decreased, and some species entirely disappeared with the advance of civilization and the increase of population." "Some," Patterson argued, "have assumed that . . . the waters were incapable of furnishing a requisite supply to meet the demands of an increasing population." Patterson rejected this assumption. "Nature," he said, "makes no such mistakes in its provisions for human wants. Filth and chemicals from various manufacturers and wastes of saw-mills, poisoning and choking the streams, the improper use of seines, pounds, and other inventions of avarice, and, finally, the erection of impassable dams . . . are the causes which have driven the fish from our inland waters." Extending Smith's idea of nature's balance, Patterson argued that the "resources of nature are all infinite. Self-destruction is the only limitation placed upon the increase of the life of the waters. . . . Human science and skill may find a practical application and help to solve [the problem of fish depletion]." Both Patterson and Smith (as did Marsh) saw the advance of civilization as causing a depletion of nature's resources. Smith believed that this decrease would be balanced somewhere beyond the frontier. Patterson saw the answer in the frontier of science. The very force that depleted nature, the advance of civilization, would provide humanity with the science and technology to restore nature. The science Patterson felt would restore fish to the waters was that of the "pisciculturists who have devoted themselves to the discovery of artificial processes of multiplication." For Americans, fish breeding, or pisciculture, which had been practiced in Europe, was a new science that might provide salvation for the problem of depleted fisheries. As Patterson noted, "This is the splendid practical triumph of science."[59]

Unfortunately, the initial reports about fish breeding were not encouraging. The Massachusetts Committee on Artificial Propagation of Fish turned to their fellow member N. E. Atwood, "a practical fisherman, and . . . a learned ichthyologist," to carry out an attempt to breed fish.[60] Although Atwood's attempt to breed trout failed, the commission nonetheless felt that his work "demonstrated its practicability."[61] In the Connecticut, the commission believed, "by means of artificial propagation, the river below Hadley Falls, might be vastly better stocked with shad than it has ever yet been."[62] The commission maintained that to initiate fish breeding, the state needed to act in a manner similar to the way it protected mills and industry. "But no legislation can be of any avail until private enterprise shall ascertain its own wants."[63] The commission concluded that "artificial propagation of fish is not only practicable but may be made very profitable, and that our fresh waters may thus be made to produce a vast amount of excellent food."[64] The commission argued that the state should encourage private fish breeding, restrict fishing to certain times and places, and investigate the possibility of more protective legislation, but nothing was done for the next eight years.[65]

By then, Vermont and New Hampshire, also concerned about the decline of migratory fish, had grown fed up with Massachusetts's failure to deal with the fish obstructions at Lawrence, Lowell, and Holyoke. Already, in 1847, New Hampshire had passed a law that required dams to be open in the spring until June 20 and to be opened in the fall by August 20 for fish migration. According to the state's commissioners on fisheries, it was "the right of New Hampshire to require these passages to be opened even if it should cause some inconvenience to mill-owners."[66] In June, 1864, Vermont and New Hampshire, realizing that nothing could be done to bring fish back to the northern reaches of the Connecticut River without action from Massachusetts and Connecticut, passed bills requesting that those states take action to do something about the obstruction of fish migration.[67] March 14–17, 1865, a Massachusetts legislative committee held hearings on the feasibility of building effective fishways on the Connecticut and Merrimack. Vermont's Governor John G. Smith sent a letter that noted that before the dam had been raised at Holyoke, "there was a great abundance of salmon, [in fact the salmon had been mostly eliminated long before the Holyoke dam], shad, and other fish in that portion of the river adjoining this state, which have since disappeared, caused by . . . obstructions . . . placed by citizens of Massachusetts." Governor Smith, under pressure from farmers and fishers along the Connecticut River "who feel aggrieved," claimed that although Vermonters did not want to "deprive the citizens of other States of the right to enjoy the privileges on the river," nonetheless they felt that "justice requir[ed]" that something be done.[68]

Governor Joseph Gilmore of New Hampshire sent Judge Henry Adams Bellows to the hearings. Bellows understood the importance of fish to rural folk in New Hampshire, particularly in the Connecticut River Valley where he had grown up. Left fatherless at an early age, Bellows had struggled with poverty and privation. Fish caught in the Connecticut had often been his family's major source of protein. After passing the bar, Bellows practiced law in the upper Connecticut River Valley for almost twenty-five years. In 1859, he was appointed to the supreme judicial court. Ten years later, he became chief justice. Bellows testified "forcibly, but with much candor," before the committee that Massachusetts's manufacturing interests could not take precedence over New Hampshire's right to have fish travel up into its waters. He argued that although "the great manufacturing interests . . . should receive all due protection, Massachusetts should not disregard the interests of her neighbors."[69]

Spokesmen for the various manufacturing interests argued that the mills needed all the waterpower they could get and that fishways would divert essential water from powering machinery.[70] They also suggested that sewage and other pollutants were even more responsible than dams for the decline of fish.[71] The legislators came away from the hearings hoping that some new fishway might restore the fish on the rivers. They also realized that something had to be done, as New Hampshire would not let the issue die.[72]

Judge Shaw's ruling in the Essex case had, in fact, put the state in a bind. The Essex Company's original charter required the county commissioners' approval of the fishway, which the commissioners had given. Shaw ruled, however, that this approval, coupled with the charter's later revision, meant that a later legislative act could not require the company to build a "sufficient" fishway. In Massachusetts, Shaw's ruling closed the issue. But New Hampshire, concerned about the loss of fish in the Connecticut as well as the Merrimack, was not satisfied. As the joint committee noted, "The rights of New Hampshire" had been injured "by the mistaken determination of the Essex commissioners, in prescribing for the fish-way at Lawrence, and ... a proper regard for the rights of others, call upon the state to rectify the error committed by its agents, if it can be done without sacrificing the greater interests which are depending upon the use of the water for manufacturing purposes."[73]

Just imagine the pressure on the local Essex commissioners in 1846 to find the Essex Company's fishway satisfactory! Here was a powerful coalition of Boston investors interested in spending millions of dollars in Essex County to build a new industrial city along the lines of the fantastically successful Lowell. All they needed was to have this local commission certify that their fishway was adequate. And, of course, the people presenting the case for the fishway were the experts from the Essex Company. Once certified as satisfactory, the Essex Company was free of obligation, whether or not fish could run up the fishway. Behind this decision, the Essex Company had Chief Justice Shaw, who not only believed that manufacturing was of significantly greater public good than fish, but who also believed that a charter was like a fixed contract and had to be protected in the interest of safe investment. The legislature was never as willing as the courts to abandon the fish, but like the courts, the legislature did not want to "sacrifice the greater interests which are depending upon the use of the water for manufacturing purposes."[74]

To protect the state's own interests, and under pressure from Vermont and New Hampshire, in May of 1865 Massachusetts created a two-person commission to investigate "the obstructions to the passage of fish in the Connecticut and Merrimack Rivers."[75] The commission was to look at the dams at Holyoke, Lowell, and Lawrence "to ascertain the extent and degree of . . . discharge of dyestuffs and other noxious matter therein from the manufactories, and the effects of such matter upon the water and the fish inhabiting the same," and to investigate fishways.[76] This river fisheries commission was also charged with determining the numbers of shad and salmon before the obstructions, and the causes of fish population decline. It was to communicate with Vermont and New Hampshire about laws concerning fish and fishways in their states.[77] In effect, the commission was being asked to bring back the state's fisheries to the satisfaction of New Hampshire and Vermont without discomforting the manufacturing interests in Massachusetts, surely a Herculean task.[78] For this Herculean role, Governor Andrew chose Alfred A. Reed of Boston and Theodore Lyman III.

In the governor's eyes, Lyman was perfect for the job. He was a recent veteran, a Harvard man, and a respected scientist and protégé of the nation's leading scientist, Louis Agassiz, who had already advised the state on the issue of fish and pisciculture. Lyman was a member of one of the leading families of the state. He could talk to all sides. If a resolution could be found for the difficulties facing the state on this issue, Lyman was the one to find it.

The governor was not the only one who believed that Lyman was perfect for the job. So did Lyman. He and Reed went right to work. Focusing on the three basic issues—obstructions, pollution, and fishways—Lyman and Reed began investigating the habits of fish and the effects of the dams at Holyoke, Lawrence, and Lowell on the fish.

In their report to the governor, Lyman and Reed noted that fish, now significantly diminished, had been plentiful in the Connecticut and Merrimack Rivers before the big dams.[79] The commissioners also noted the importance of fish as food for the region's inhabitants.[80] Concerning fishways and the power needed by the mills, the commissioners noted that there was enough "spare" water for a fishway at Lawrence during the spring fish run, and that there was "plenty of water at Holyoke."[81] Lowell proved to be more difficult for the commissioners. Noting the testimony of James Francis, the Lowell engineer, before the joint legislative Commission on Obstructions to the Passage of Fish in the Connecticut and Merrimack Rivers, the commissioners admitted that "a scarcity of water at [Lowell] is a serious matter. It not only touches the capitalist, but immediately affects between eleven and twelve thousand operatives. If, from slack water, the machinery runs slow, those operatives who work by the piece earn little and are discontented."[82] Yet the commissioners came to the conclusion that it was possible to build a fishway without greatly harming the operations at Lowell. Pollution proved a harder issue.[83]

Pollution and Fish

Wastes, both human sewage and industrial byproducts, affect the water system in several ways. Live water systems contain dissolved oxygen (DO). Although not understood for most of the nineteenth century, this dissolved oxygen supports marine life and organisms that function as the cleanup crew. Organic matter within the water system breaks down by interacting with the oxygen in the water. Live plant life and the movement of water capture oxygen from the air to contribute to the flow of oxygen into the system. This process enriches aquatic plant life, which provides the food for fish and smaller marine life. A healthy water system has a high level of dissolved oxygen. When high levels of nitrates and phosphates are poured into a water system, they generate plant growth, which dies and can overwhelm the oxygen level in the system. Fish need relatively high levels of dissolved oxygen to live. So when nitrates flood into the system and the DO level drops, fish die out. Sewage also absorbs oxygen as it breaks down. Sewage in water systems represents a biochemical oxygen demand (BOD) that reduces the volume of dissolved oxygen in the water.

High BOD wastes were not the only problems for the waters of New England. New England industries also dumped waste chemicals that directly killed aquatic life, including the bacteria that help break down pollutants. Solids flushed from industry vats settled on the bottom of streams and rivers, smothering bottom-living marine life and creating poisonous sediments. Although nineteenth-century New Englanders did not completely understand the various processes by which industrial wastes, particularly chemical wastes, disrupted and destroyed water systems, many of them understood that industrial wastes corrupted water systems and needed to be controlled.

As a scientist, Lyman did not want to generalize from impressions. He preferred the sanctity of experiments and scientific results. He noted that pollution from the mills was clearly visible. "There is a rush from the raceways of inky-liquid, which, at Lawrence, may often be distinctly traced three hundred or four hundred feet across the stream . . . while a curdy froth of soapsuds floats down the river, and may be seen in quantity more than a mile below the town."[84] Yet for Lyman "to state in a comprehensive way what is the effect of certain impurities in water, [was by] no means easy." "Science needed to be exact."[85] To do an adequate job of providing the scientific evidence to make an informed decision required massive experimentation, which the commissioners felt was outside their power. Instead, they conducted six tests to examine the impact of certain pollutants on live fish. From these tests, the commission noted that in most cases lime, acids, and heavy concentrations of soap killed the fish. The two commissioners also observed that the gasworks at Holyoke dumped surplus tar, coal oils, and ammonia into the water, which killed the shad and made other fish inedible. City sewage did not seem to be as fatal. These experiments and observations convinced the commission that pollution was harmful, but they could not "prove whether these factory pollutions actually do drive fish from a river"[86]

In the eyes of the positivist and improvement-oriented Lyman, pollution was a real problem not only because there seemed to be evidence suggesting its negative impact on fish but also because it was unnecessary and wasteful. Paper mills and print-and-dye works that dumped lime into rivers not only killed the waters but also threw out "a valuable manure." Lyman saw these rivers in mechanistic terms. They had multiple uses, but like all machines, they needed to be used but not abused. Dumping reusable wastes into the river was "simply slovenly." "In a word," Lyman noted, "a fair stream is a mechanical power and a lavatory, but it is not a common sewer."[87] If manufacturers were more conscientious in their disposal of wastes, he reasoned, the effects of the remaining pollution, if controlled (for example, "confined to one side of the river"), could be mitigated. This would be especially so considering "the well-known power of rivers to work themselves clean."[88]

Fishways

The commissioners concluded that for a successful return of shad and salmon to the Connecticut and Merrimack, not only should pollution be moderated but fishways needed to be built around the dams. They could be built to allow both the passage of fish and the profitable running of the mills (despite the inconvenience to the manufacturers). The commissioners also argued for the passage of stringent laws regulating fishing in all of the affected states, and for shad and salmon to be bred.[89]

Lyman and Reed reviewed the law and the court decisions on fishways. Lyman noted in their report, "The state, in the exercise of the right of eminent domain, may construct such fish-ways as may be deemed expedient."[90]

The commission's original charge was not only to report to Massachusetts but also to work with the commissioners from New Hampshire and Vermont.[91] In June 1865, while Lyman and Reed were still investigating for Massachusetts, New Hampshire passed a stringent antidam law. In New Hampshire, the mill owners did not have the kind of clout the Boston Associates exercised in Massachusetts. The New Hampshire law also created a permanent commission on fisheries "to go into effect whenever

... suitable fish-ways for passage of sea-fish over the dams on said river or rivers, below the boundary of this state shall have been commenced."[92] New Hampshire's fish commissioner, Judge Henry Adams Bellows, was to consider the subject of restoring anadromous fish to the Connecticut and Merrimack as well as the subject of introducing new varieties of freshwater fish. Vermont founded its fish commission in 1865 and also pressured Massachusetts and Connecticut to do something to restore fish to the Connecticut River.[93]

After turning in his report, Lyman went to Lawrence to meet with representatives of the other New England states. It was at this meeting that New Hampshire governor Smyth threatened to take stronger action if Massachusetts did not do something soon. New Hampshire and Vermont both pressed Lyman for action.[94]

In May of 1866, under pressure from New Hampshire and Vermont, Massachusetts passed a fish act "concerning the Obstructions to the Passage of Fish in the Connecticut and Merrimack Rivers." The act created a permanent fish commission to begin planning for fishways and to work with the other states on these plans. The commission was also charged with gaining consent to the plans from the proprietors of dams and with ensuring the proprietors maintained the fishways to the commissioners' satisfaction. The act also called on the State of Connecticut to appoint a fish commission.[95] The State of Connecticut responded in 1866 by establishing its fish commission, with F. W. Russell and Henry Robinson as commissioners.

On July 3, 1866, the governor appointed Lyman one of the permanent commissioners of the Commission on Inland Fisheries and sent him, "that very day," to Concord, New Hampshire, to convince Governor Smyth of Massachusetts's sincerity.[96] Lyman was able to persuade New Hampshire that Massachusetts was determined to take action, and within ten days, he met with mill officials at Lawrence and Lowell to discuss fishway construction.[97] By the end of 1866, the Massachusetts commission had plans for fishways at Lowell and Lawrence.[98] Lyman, like Patterson, believed that science and technology could mediate progress. Fishways, incorporating the latest design and technology, could open a passage for fish. But it was the science of pisciculture, Lyman believed, that would restore the abundance of the region's great rivers. Following the commission's recommendation, Massachusetts set up the nation's first fish hatchery, on the Connecticut at Hadley Falls. To run the hatchery, Massachusetts hired Seth Green of New York, the nation's leading expert on fish breeding. Green's hatchery soon became a source of millions of shad fry, which were released into the Connecticut and brought a dramatic increase in the number of shad in the river.

In February of 1867, the fish commissioners from Vermont, New Hampshire, and Connecticut met with Lyman and Alfred Field to coordinate activity for action on the Connecticut River. At the first official meeting, a dinner meeting at Lyman's Brookline home, Judge Bellows of New Hampshire was elected chair of the New England Commission of River Fisheries and Lyman secretary.[99] The New England commission's task involved Massachusetts building fishways and restricting fishing; Vermont, New Hampshire, and Massachusetts increasing fish breeding; and Connecticut controlling the use of gill nets and weirs on the lower Connecticut River.

Lyman was pleased with his accomplishments and felt comfortable with his fellow commissioners, even the insistent Judge Bellows. Most of the members of the various state commissions shared Lyman's belief in progress and science. Most also shared his paternalism, but the Connecticut fish commissioners did not. Lyman did not care for them. He viewed one as "half educated" and the other as being a self-interested gillnetter. Lyman saw himself as a disinterested scientist. Although he held significant amounts of stock in the mills whose dams and pollution were hurting the fisheries, he could not see his own self-interest. Yet he perceived the Connecticut commissioners as "unfit" because they lacked his scientific training and objectivity.[100]

Besides his other fish commission work, Lyman also discussed the fish situation with mill officials at Holyoke, Lowell, and Lawrence.[101] The meetings with the officials at Lowell and Lawrence were comfortable for Lyman. The officials of the textile companies and dams came from the same social community he did. Lyman knew them or knew of them. At Lowell, "Mr. Francis [the Boston Associates superintendent] treated us with the usual hospitality."[102] During another visit to Lowell, the "good Mr. Francis," after showing off the progress on the fishway, brought Lyman home for tea, peaches, and plums with Mrs. Francis. Before leaving the city, Lyman had dinner with James Thurston, another mill supervisor.[103]

Lyman believed that with the cooperation of the mill owners and with the aid of science and technology, fish could be returned to the rivers of New England. He assumed that the mill owners would cooperate because he knew them, shared tea with them, had gone to school with them, and held stock in their mills. The mill owners were more than willing to have tea with Lyman; they were, after all, from the same social set. They offered Lyman cooperation, but only on the terms that Lyman's science and technology not interfere with their profits. This Lyman not only understood but agreed with.

Lyman sought a happy accommodation that would allow for the steady progress of science and technology, maximum profits for the mill owners, and also abundant fish runs, so that "a poor man with his quarter may obtain a good meal."[104] To accomplish this, Lyman felt that a stronger fish act was required. The act Lyman wanted would increase the power of the state to regulate the fisheries and control water pollution, while equating fishing to an industry. To do so, Lyman believed, fish had to be seen as property, and fishers given property rights over free-running fish. For Lyman, the restoration of fish as food for the poor did not mean the free fishing of old, but "cheap fish" produced for profit.

In Lyman's 1868 report to the state, he argued that a private fish industry could return a profit of two hundred to one, "and yet our clever people go on year after year putting up more thousands of spindles and flooding the market with unusable cotton goods when from the very water which turns their wheels, they might coin money with no other machinery than a net and a hatching trough."[105] Lyman believed that sensible people, including industrialists, would see that the market indicated that restoring the region's fisheries was a good investment. Fisheries and industry did not have to be in conflict, since as Lyman noted, the textile market did not need more cloth, while the market for fish was expanding. This view was not shared by the textile companies and indeed was a surprising utterance from Lyman himself.

Fish as Food

Although George Perkins Marsh noted the importance of fishing as a recreation that reinforced the manly virtues, and sportsmen fishers like Lyman and the members of the New Hampshire Game and Fish League avidly took to the rivers and streams to catch fish for sport, it was not until the end of the century that profishing advocates made the claim that fish should be restored for sport. For nineteenth-century reformers, fish restoration had two goals: to supply food to the poor and to provide a profit to fish farmers.[106] Patterson reminded the members of the New Hampshire Game and Fish League that "if game protection and fish culture are simply to furnish amusement to the idle, if they can only supply an occasional dainty to the tables of the rich . . . we shall fail as we ought, to exalt them into a public interest, but if they can be made sources of general income and add largely to the food-supply of our increasing population, then they may demand popular patronage and draw legitimately from state and national treasuries for their development."[107] Quoting from Professor Spencer F. Baird of the recently created U.S. Commission on Fisheries, Patterson noted that "water may be made to produce as much in value per acre as land."[108]

In 1866, when New Hampshire fish commissioners were petitioning for strong fish laws and restoration, they also argued that bringing back fish "would furnish a very large supply of food at a small expense."[109] In 1871,

New Hampshire noted that "the time is fast approaching when our people of moderate means must find it difficult to procure wholesome food in sufficient quantity to sustain themselves and their families—as was the case in olden times when this state was first settled and when they were obliged to resort to the waters for a large share of their subsistence—and there can be no question that providing fish for such an emergency is the cheapest and the best and the most human step that can be taken for the benefit of the greater number."[110]

The argument for fish restoration played upon popular romanticized memories of a more idyllic time, when people were more able to live off the land, even if these memories were mellowed with the passage of time. Yet that image had another side, that of the lazy ne'er-do-well, unwilling to work because of the abundance of nature. In the age of industry and progress, the fish commissioners were careful not to let the vision of nature's abundance feed into this second image. Instead, they tried to link an increase of nature's abundance to a corresponding increase in enterprise.

> Some have objected to the enterprise [of stocking fish] on the ground that the more fish we have the more we should encourage a set of lazy fellows to fish who are too lazy to do anything but fish. Now, suppose we admit that to be a fact, it is an argument in favor of the enterprise for we all know that these lazy fellows catch nothing now whereas if fish were plenty they would catch fish for themselves and others to eat, and would there by keep themselves from our poor farms. If the number of the lazy who will not work, but will fish is half as large as some would have us believe would it not be wise for the state to create an industry which this large body of its citizens might with certainty benefit themselves and the whole community.[111]

Fish as Property: A Limit on Popular Rights

In his 1869 report to a joint committee of the legislature, Lyman submitted a proposal for a new fish act. The new act dealt with several issues. Lyman asked for stronger legislation for fishways and against the polluting of streams, rivers, and ponds. He requested restrictions on when and how fish could be caught and suggested that enforcement power be put into the hands of the commissioners. Finally, he asked that the ancient charters' rights be amended.[112]

The ancient charters, as noted earlier, allowed any citizen to pass and repass on foot through any person's property to fish and fowl on any great pond larger than ten acres.[113] Arguing from his own experience as an improver of his farm and a beneficiary of the abundance of private capitalism, Lyman argued that ultimately, the only remedy for the diminishing numbers of fish was "to make fishes under certain conditions, property and thus give the same stimulus to the cultivation of fish as that of other live stock."[114] Lyman felt that the "need for such a law ... is the same as the need for a law protecting our industry."[115]

By revising the ancient charters so that ponds of between ten and twenty acres would not be open for fishing and fowling, public ponds could be leased, and proprietorship over the fish in these ponds was given to the lessee, Lyman hoped that the same spirit that encouraged husbandry and industry among farmers and industrialists would infuse the fish farmers. He called for the government to privatize public rights into the hands of small property-owning farmers. "Certainly they [the farmers] would not hesitate to begin this industry did they know the little labor needed and the considerable food and profit to be derived."[116] Lest any of the legislators be concerned about the loss of customary rights to fishing and fowling, Lyman assured them that although any citizen could pass over another's property, "no self-respecting gentleman would exercise his right to trespass on property of a farmer to gain access to a pond of ten acres."[117]

Despite Lyman's argument that no self-respecting gentleman would trespass, he conceded that poorer folk were more inclined to take advantage of this right. As a member of the New England elite, Lyman was suspicious of the motives and actions of the poor. While he believed that gentlemen of science could and would protect a resource, the poor were, in his mind, more likely to give in to their immediate wants, even if doing so would produce negative results. Without government supervision or property rights over the resources, he felt the poorer sort were liable to overuse the resource and destroy it.[118]

Proof of this was found in the experience of Lyman's friend Samuel Tisdale. Tisdale began stocking his ponds in the 1850s. "So soon as the fact was known, all the neighborhood at once gave its assiduous attention to poaching—indignant that anyone should be so aristocratic as to try and furnish cheap food to the community. Their efforts were so far successful as to much reduce the number of fish."[119]

Lyman's efforts for a new fisheries law proved successful. In 1869, Massachusetts passed the Act for Encouraging the Cultivation of Useful Fishes, which created legislation for the empowerment of the commission's restrictions on when people could fish and how they could fish, encouragement of fish breeding, and the revision of customary rights to fish and fowl, which limited public access to ponds of less than twenty acres (as opposed to less than ten) and allowed the commissioners to lease public waters to private individuals. This act gave the commissioners not only power to enforce the building of fishways but also the general management of fish resources for the state.[120]

The provision Lyman wanted in the new act concerning pollution was not particularly strong. He proposed a law stating that "after first notice, anyone who corrupts the waters of a pond, or stream by placing or suffering to flow therein, any drugs, dye-stuff, lime, tannery, liquors, or other substances injurious to fishes, shall for every day... corruption continues forfeit fifty dollars." But the law would also provide "that the owners of manufactories now existing whose processes require the discharge of said injurious substances, may allow them to pass into the waters of such ponds or streams when the commissioners of inland fisheries or either of them certifies that such substances can be disposed of in no other reasonable way."[121] As reasonable and cooperative as Lyman felt this proposal was, the manufacturers saw it as a threat, and they fought it in the legislature. The mill owners did not want commissioners deciding what was reasonable, even if they were backed by scientific expertise. The manufacturers had any provisions on pollution removed from the final version of the new fisheries act of 1869. Lyman was disappointed with this but not disheartened. Although he suspected that industrial pollution was harmful to fish, he did not believe that the evidence at hand indicated pollution was as significant a problem as the dam obstructions. And the fish act did address the obstructions.

Holyoke Fights Back

Passage of the fish act did not end Lyman's difficulties. Although the mill owners at Lowell and Lawrence were willing to cooperate on building a fishway, those at Holyoke were more resistant. Following passage of the fish acts of 1866 and 1869, the Holyoke Waterpower Company "utterly refuse[d] to construct said fishway, or to agree... for the construction of the same."[122] Lyman's hope that he could persuade the Holyoke people to use the latest science and technology to harmonize the conflict between the needs of the migratory fish and the mills proved too optimistic. Reluctantly, Lyman went to court.[123] Before the court, Lyman claimed that the company had an obligation under the law to allow the building of a fishway. He explained that the amount of shad in the Connecticut had fallen 25 percent since the building of the dam. While the Holyoke company admitted that shad had been cut off from above the dam, they claimed to have paid damages to those with fishing rights above it. The company denied that the dam had any impact on the shad below the dam. They blamed overfishing, the clearing of woods, and the dumping of sewerage and manufacturing pollutants into the river for the decline. The dam owners claimed that because of these factors, the number of fish would have declined even without the dam. Moreover, the company argued "that there are now more than twenty different corporations dependent upon said waterpower to carry on their business . . . involving 96,000 spindles, 24,000 sets of woollen machines, 117 paper engines . . . and other machinery, requiring and controlling several millions of dollars of capital and gathering to it, and furnishing business for a growing town."[124] The Holyoke Waterpower Company claimed that if a fishway was built, power would be diverted, and as a result, dividends to stockholders would have to be withheld. The company felt that it had an obligation to its stockholders not to build the fishway.[125]

Lyman won a surprising victory before the Massachusetts court, but the Holyoke Company appealed to the Supreme Court of the United States.[126] The case came before the Court in 1872, with Justice Nathan Clifford delivering the opinion of the Court. In his decision, Clifford surpassed what the courts in Massachusetts were willing to say. He argued that rivers and streams "may be regarded as public rights, subject to legislative control," and "fisheries even in waters not navigable, are also so far public rights." Although Clifford noted that the issue of riparian rights in other states was not addressed in this particular case, he cast doubt on whether the rights Chief Justice Shaw gave the Essex Company under its Massachusetts charter could be maintained on a river that ran through two or more states. Even without that issue, Clifford argued that the corporation was not immune from legislative requirements to build a sufficient fishway.[127] The Massachusetts Commission on Inland Fisheries won again.

Emboldened by the strength of the Supreme Court's decision, the New England commissioners aggressively worked to have fishways built around the major dams. They continued to work with fish breeding and hatcheries and to encourage conservation. In 1876, Lyman was called to Washington to chair the national conference of U.S. fish commissioners.[128]

But with all the commission's activity, science, and technology, the goal of returning anadromous fish to the Connecticut remained elusive.[129] Seth Green's shad hatchery did, indeed, lead to renewed prolific runs of shad up the Connecticut in the 1870s and 1880s. But problems with the fishway at Holyoke discouraged cooperative efforts among the states. The State of Connecticut, under pressure from their fishers, opened up the restrictions on fishing on the lower Connecticut, which enraged the Massachusetts commissioners. Vermont and New Hampshire were likewise discouraged by the failure of the fishway at Holyoke and gave up their shad-hatching efforts.

If Lyman's attempt to overcome the conflicting interests of the manufacturers and the fish and ecology of the river systems fell short, we should not be surprised. From today's perspective, Lyman's faith that science and technology coupled with sincerity and an ability to sit down and discuss the issues like gentlemen and reasonable people seems naive to an extreme. But we must remember that Lyman lived in the midst of the positivist age. He not only believed in science and technology, he felt he had seen it succeed on his own farm and at the farm school. As a successful investor, Lyman also believed in the market and considered "Americans [to be] a business people." If profits could encourage manufacturers to turn New England from "nothing but a few gristmills" into a prosperous industrial success story, so too could profits encourage the development of a fish industry to bring back the old abundance. With the aid of science and technology, Lyman hoped to overcome the problems of the obstructions in the way of fish migration. Through pisciculture, fish stocking, and the privatization of what had once been seen as a public resource, he hoped to bring the profit motive to bear on the raising of fish. Because Lyman was of the same class as those who ran the mills and operated the dams, he assumed that he could sit down with them and work out an agreement to the benefit of everyone. As Lyman himself noted, "The commissioners felt it to be part of their duty to harmonize, as far as possible, the conflict that had existed for more than a century between the public rights and the private interests of the mill owners."[130] Moderation, science, and service were Lyman's creed his whole life. He saw no reason that these would fail him in this most important task. Of course, they did. On the big dams, the fishways Lyman was so proud of proved to be failures. It was later remarked by a Connecticut Commission of Fisheries and Game that "a healthy and badly scared red squire, if the way was free of water, might get up [the fishway] but from the knowledge at hand, no fish ever succeeded in doing so, nor is likely to, till the breeds are materially improved as to their power of locomotion."[131]

The work of Lyman and his commissioners did not end with the failed fishways. The pattern established by Lyman of science, technology, privatism, and cooperation continued into the reforms of the late nineteenth century and into the twentieth. Indeed, they were the hallmarks of American Progressivism. But before categorizing Lyman and his fish commissioners as failures, we should remember more than simply their shortcomings. Lyman may not have accomplished what he had hoped—in fact, it is even doubtful that he realized how little he actually did accomplish—but perhaps more significant is that the attempt was even made at all.

The manufacturing interests of Massachusetts represented a mammoth consolidation of wealth and power. Those few families who made up the Boston Associates held in their hands unimagined wealth. They owned banks, insurance companies, railroads, mills, machine companies, and real estate. They and their relatives made law as judges, controlled universities, and held leading professorships. They periodically occupied the various high political offices of the state. They controlled the payrolls of thousands of workers and even whole towns. Yet with all their power and wealth, the Boston Associates and their cohorts were not able to prevent common fishers and farmers from continually demanding that something be done to address the needs of the fish. Despite tremendous pressure from the corporations to ignore the fish, the legislature continued right through to the last quarter of the nineteenth century to pass legislation that attempted to protect the fish. As the Vermont Commissioners of Fisheries noted in 1867, "Public opinion is growing stronger in favor of [fish laws]."[132] Henry David Thoreau asked in 1839, "Who hears the fishes when they cry?" The answer, in a strange sense, was more people than one might think, given the forces and voices attempting to drown out their piscine entreaties.

6

Health, State Medicine, and Henry Ingersoll Bowditch

The Radical Approach

State Board of Health

On September 15, 1869, Massachusetts governor Andrew appointed seven members to the state board of health.[1] The men appointed to that board had a new vision of medicine and the roles of science and the state in protecting health. For these men, medicine should do more than just cure; it must also prevent illness. Their understanding of illness was expansive, and it led them to a concern about filth and pollution. They also came to believe that for science and medicine to perform their new role in society, they needed the backing and power of the state.

On September 22, the board met for the first time, electing George Derby as secretary and Henry Ingersoll Bowditch as chair. Bowditch was a logical choice for chair. In addition to being one of the region's leading doctors, he came from a respected Boston family, and he held the professorship of clinical medicine at Harvard School of Medicine.[2] He was vice president of the American Medical Association (later he would be president) and the author of several scientific journal articles. Bowditch served as a medical volunteer to the Union army and lost a son in battle.[3] Moreover, it had been his idea to form a state board of health.

In a speech before the Massachusetts Medical Society in 1862, Bowditch argued that medicine should serve the people. To do so required the creation of a state board of health, "one that eventually will be of more service . . . to the inhabitants of this state . . . by [its] united and persistent efforts to increase the state authority."[4] Bowditch was not the only one to advocate for a state board.[5] Dr. Edward Jarvis, a well known sanitary reformer, had as well, and along with Bowditch, he pushed the idea, only to have it fail in the legislative house in April of 1866 as "inexpedient," despite Governor Andrew's endorsement.[6]

Three years later, a typhoid epidemic in western Massachusetts encouraged state representatives from the Connecticut River Valley and farther west to back a bill for a state board. The wife of Democratic Party boss Thomas Plunkett read Bowditch's 1866 speech and persuaded her husband to sponsor a bill to create the board.[7] With her encouragement, Plunkett pushed the bill in the 1869 legislature with the argument that "it is cheaper to try to prevent rather than fail to cure a disease."[8] With both Democratic and Republican support, the bill passed.[9]

Henry Ingersoll Bowditch

On paper, Bowditch's selection to head the Massachusetts State Board of Health—the first such board in the nation—made logical sense. Upon closer examination, he was also an interesting choice. More than a staid Boston Brahmin doctor, Bowditch was also a self-proclaimed "radical." Indeed, as he confessed to his old childhood minister, "I seemed called to do battle for justice and the oppressed."[10] And do battle he did. Over the preceding three decades, Bowditch had forged a reputation as an uncompromising militant abolitionist and defender of the outcast.[11] He joined the radical wing of the abolitionist movement in the 1830s and remained with it until the end of the war. After the passage of the Fugitive Slave Law, Bowditch became involved in both successful and unsuccessful runaway slave rescues. He was vocal and visible at antislavery demonstrations, rallies, and gatherings. He was an open friend of Frederick Douglass (a guest at Bowditch's home) and of William Lloyd Garrison. He sat on the board of the New England Anti-slavery Society.[12] After a protest against the enforcement of the Fugitive Slave Law, the *Boston Courier* wrote an editorial denouncing Bowditch: "If a physician is eagerly running about town to help break the laws, if we hear of his offering money to a jailor to let one of his prisoners go free; if he is secretary of noisy political meetings, if he makes speeches in the streets, we do not ask him to come and see us when we are sick."[13]

In addition to being a radical abolitionist, Bowditch also described himself as a "decided Women's Rights Man," a stance perhaps even less popular than his abolitionism.[14] As vice president of the AMA in 1867, Bowditch antagonized the organization by his outspoken "views [in support of] women Doctors." Bowditch claimed he "couldn't help" supporting the unpopular idea of women's rights, for he could not "sit by and see an honest cause abused and spit upon without at least protesting."[15] Bowditch also championed the poor.[16]

Although he believed his radicalization stemmed from seeing a crowd attempt to tar and feather William Lloyd Garrison, in fact, the stage for this revelation was built over his years growing up in a liberal home and during his two years of study in France.[17] Bowditch was born in Salem in

1808. When he was fifteen, the family moved to Boston. Bowditch graduated from Harvard in 1828. Not wanting a career in business, law, or the ministry, he entered Harvard Medical School, more by default than for any other reason. After Bowditch finished medical school and an internship at Massachusetts General Hospital, his father agreed to send him to Europe to complete his medical education. It was not unusual for American scholars like Bowditch and Theodore Lyman, interested in furthering their academic work, to study abroad. Bowditch began his studies in Paris at the Ecole de Medicine, leaving for Europe in 1832. Once he reached Paris, his father's connections facilitated Bowditch's acceptance into the community of scholars and academics. He also quickly fell in with fellow New Englanders who were also studying in Paris. Oliver Wendell Holmes, Mason Warren, and Copley Green lived nearby. With them, he spent hours talking art, philosophy, and politics in cafes and one another's rooms.[18] If Bowditch entered medicine by default, his time in Paris awakened a love of it. Under Dr. P. C. A. Louis's tutelage, he came to appreciate medicine as a profession of caring and intellectual challenge.[19]

In Paris, Bowditch gained more than just a love of medicine. He also discovered politics. By his second year there, Bowditch's comments on the government of Louis Philippe became more partisan and critical, and he became involved with French thinkers and republican activists. In anger, he wrote his parents that the government had turned its back on the revolution; "so much for the Government of the Citizen King!"[20] Paris in the 1830s was an exciting place to be, and its excitement infected Bowditch. "I am far more liberal now than I was before visiting Europe," he wrote to his mother in February of 1834.[21]

Bowditch's liberalism was not something he planned to leave behind in Paris. "I want to see everything more free than it is now. . . . My grand aim shall be to give to everybody the opportunity to study, and in this way repay to humanity at large the immense debt of gratitude I owe to France."[22]

More than politics and medicine caught Bowditch's fancy in Paris. Upon returning to the city in the fall of 1833 to continue his second year of study under Louis, Bowditch moved from his old lodgings to a room in a *pension bourgeoise* at number 1, Rue d' Aubenton. In the *pension* also lived a seventeen-year-old English woman completing her education in Paris, and her elderly aunt. The young woman was Olivia Yardley, and Henry fell in love with her. Four years later on the seventeenth of July 1838, only two days after Olivia arrived in America, the two were married.[23]

Upon his return to Boston, Bowditch threw himself into developing his practice. In the midst of his attempt to set up his practice, Bowditch witnessed the mob attack on Garrison. Bowditch tells us that this was the moment when he committed himself to the cause that would become the driving force in his life for the next thirty years.

Despite his steadfast refusal to sacrifice his principles and antislavery activities in the face of stern opposition from colleagues in the medical community, his work as a doctor continued to gain him fame and position.[24] As his position grew more secure, Bowditch had the time to take vacations with his family. He took his sons and assorted friends and relatives on summer wilderness excursions, first up the Penobscot in birchbark canoes, and then into the Adirondacks. He found peace in the wilderness, "a solemn silence," which made him thankful for the richness of the wilderness experience.[25] While camping in 1866, Bowditch rejoiced in his journal for the calming influence of nature. "The woods! The woods! They are the elixir of life for me." Bowditch did not look to the woods as a place of "sport" but "for communion with nature."[26]

Bowditch also liked to combine a retreat from the burdens of work and political agitation with science. In 1849, Lyman's teacher, Louis Agassiz, invited Bowditch on a scientific excursion off the coast of Cape Cod to dredge marine samples from the ocean floor. The experience was exhilarating for Bowditch: "My whole being harmonizes with nature: perfect health, a clear, transparent, balmy atmosphere, and pure scientific research."[27]

Although Bowditch expended most of his political energy on abolitionism, he also struggled to improve the overall conditions of the poor. Bowditch believed that the poor needed improved housing. "I was determined that something ought to be done about improving the tenements of the poor."[28]

By volunteering both his knowledge and his medical expertise to the war effort, Bowditch proved his commitment to serving Massachusetts faithfully. Yet he had also demonstrated a dedication to principle, commitment to the poor, reluctance to compromise, and lack of deference to—indeed an outright suspicion of—the state's leading families. It was because of these other combative qualities that the governor was taking a risk by naming Bowditch to the first board of health.

Bowditch's commitment to "do battle for justice and the oppressed" defined his radicalism. His radicalism decried not only inequality before the law but also inequality in practice. Bowditch understood that people had interests and privileges and worked to protect them. He also believed that the battle for justice involved confronting those people and their interests and privileges. Bowditch did not limit his concern for the oppressed to the faraway slave but brought it to bear on issues of women's rights and the poor. Bowditch's radicalism was defined not only by his willingness to oppose injustice to the poor and oppressed but also in his willingness to attack privilege, particularly economic privilege.[29]

Once he was on the board, Bowditch surprised no one by being elected chair. Three other members of the board had also been active abolitionists, including Bowditch's two fellow doctors, Robert Davis and George Derby. Like Bowditch, Davis had participated in the radical wing of the movement. Derby had also been an early supporter of a state board of health and had been editor of the state's vital statistics publications.[30] These former abolitionists brought their political awareness and a commitment to struggle to the campaign for public health.[31]

The Reform Impetus and the Problem of Pollution

Concern for public health arose as New England left its rural past and entered an industrial future. As manufacturing towns and cities expanded, the problems of pollution and "pestiferous centers of disease and death [once thought] . . . peculiar to the old cities of Europe, [were found] here in this newer country."[32] Increasing rates of morbidity and mortality in the cities of Massachusetts had been documented for almost two decades. With the war behind it, the state now mobilized to do something about them. Unlike the earlier health reformers who saw unhealthy conditions as unnatural, contrary to God's design, and a product of ignorance and poor habits, these postwar reformers saw public health in terms of society's responsibility.[33]

Health reformers had no trouble experiencing environmental problems first hand. They could see, taste, and smell the pollution all around them. In this age of sensual empiricism, in which personal experience dictated understanding, the fact that the air was dirty was as evident as the smoke that hung over the cities. Dirty water was "turbid," "rank," and "foul tasting."[34] For these reformers, the language of pollution was the language of the senses. Air and water were polluted if they were unpleasant to the eye, nose, or tongue. In 1885, in a report on the Hockanum River, which ran into the Connecticut River south of Hartford, an inspector for the Connecticut State Board of Health noted, "The effect of the mills' pollution is very apparent in the discolored water, and in the lodgement of discolored foam near the race dam. I learn of some complaints of people along the river below based chiefly on the discolored water."[35] When the Connecticut board of health official investigated the pollution of Piper's Brook and the Park River, he noted that the river above the Stanley Works was "clear, and to all appearances, pure . . . [but downriver it] ha[d] a distinct, but not very mawkish sewer odor . . . [it had] a thick dishwater appearance, filled with organic matter in abundance, with numerous fragments of feces . . . at times [it was] much darker in color, due doubtless, to the manufactory discharge." As he walked downstream, he noted that "feces are readily traced for a couple of miles, gradually becoming more comminuted and the water of a more homogeneous whitish color. At the confluence of Luther's brook, the pure clear water forms a very conspicuous line of demarcation for a considerable distance. At Newington the water had become clearer, but still is very visibly contaminated . . . [with] human feces." Despite the possible presence of feces, the investigator "muster[ed] sufficient resolution to taste the water which I did several times."[36] In its first year of existence, the New Hampshire State Board of Health investigated polluted water by looking and by smelling water samples. One sample that the board decided definitely represented polluted water "gave of an intolerable stench so offensive the sample had to be placed outside the building."[37]

Environmentalists' Theory of Disease Causation

Polluted water was a problem for health reformers because it smelled and gave off bad air. Cleaning up polluted water, they believed, would end miasma and reduce disease. The building of sewers to flush the filth of sewage away from living areas and the draining of marshes and swamps did reduce the mortality in urban areas, but not for the reasons these anticontagionists believed. Yet the reduced mortality increased their credibility. At a time before the scientific works of Joseph Lister, Louis Pasteur, Jacob Henle, and Robert Koch were commonly accepted in this country, most people believed that water naturally cleaned itself through dilution.[38] "Those who thus dispose of it," the Massachusetts State Board of Health noted in 1872, "if they think at all about its final destination, hope that the stream will purify it in some way so that it will be no longer a nuisance or offense."[39] Most nineteenth-century Americans believed that all that was needed was enough clean water to eliminate the smell through dilution. When Hartford built its sewer system in the 1840s and 1850s, city leaders assumed that the sewage would be taken care of by "diluting" it with Connecticut River water.[40] Yet visual and olfactory evidence indicated that many of the region's waters were not cleaning themselves.

By the end of the 1870s, more and more New Englanders realized "that most rivers and streams [were] polluted by a discharge into them of crude sewage, which practice is highly objectionable."[41] The water of smaller brooks like the Hop or Piper's was not only unsafe to drink but "discolored" and "turbid," and it gave off "foul offensive odors."[42]

As local brooks and streams became overloaded with sewage and, like Piper's Brook or Town Brook Sewer, became themselves sewers, towns were forced to enlarge and expand their sewer systems. As the New Hampshire State Board of Health noted in 1884, "When the need of a complete sewerage system is generally acknowledged in any village, it is usually an indication that its natural watercourses are already made foul by sink and excrementitious drainage."[43] But "the complete system" most towns and cities built in the second half of the nineteenth century consisted of either culverts that enclosed streams and made them part of the existing sewer system, or interceptor lines that picked up the sewage from the lines that had earlier dumped it into the nearest point on the river or stream and moved it away from the city center, only to dump it downstream.[44] Even as these solutions were being put into practice, residents understood that they would ultimately fail to deal with the problem. As the Massachusetts Board of Health warned, "Thus while the crowded population is relieving itself, effectively and economically of its refuse and waste material, it is turning them over in the shape of defiled water, to the injury and abridgement of the rights of every riparian owner."[45] New Englanders realized that interceptor lines and longer sewers were not the answer.[46] As early as 1872, the Massachusetts State Board of Health stated that "unless some practicable means be found of diverting . . . filth from our currents," "even our largest rivers ultimately [will] become sewers."[47]

People saw the pollution of rivers as "a growing evil," not just because it turned once pristine rivers and streams into "open cesspools" but also because of the link between polluted water and poor health.[48] Typhoid, dysentery, diarrhea, and cholera took a terrible toll on the residents of nineteenth-century cities, especially the young, and as the century progressed and more New Englanders made their home in urban areas, concern over pollution and disease grew.[49] In 1876, the Massachusetts State Board of Health found typhoid and meningitis related to pollution from factory privies in Chicopee, an epidemic of typhoid in Holyoke related to the town's cesspool, typhoid in Lawrence, and diphtheria, dysentery, diarrhea, meningitis, and cholera in towns throughout the state.[50] In 1890 in Springfield alone, fourteen people died of typhoid fever, while another thirty-two typhoid victims died that year in Hartford, followed by forty-one the next.[51]

Common citizens as well as doctors raised concerns about pollution of the region's waterways. A farmer, Mr. Hurlburt, complained to the Connecticut Board of Health that another farmer near him on Piper's Brook was taken sick and died of "typho-malarial fever," while another man "in his employ also sickened and died of the same disease. Still another man living near the brook was taken sick of the same fever." Farmers living downstream from New Britain complained that "the water is rendered wholly unfit for the use of cattle." Mr. L. S. Wells, who lived two miles downstream from New Britain, complained of "a distinct odor in summer season as well as winter" and said his cows had been getting sick after drinking the water. Mr. Francis of Newington also complained of the smell and said his cows wouldn't drink the water. His neighbor Stoddard complained that the "very offensive" smell of the polluted water was so strong that it reached his house. Another resident along the stream objected to the odor of the stream, which used to be clear and filled with fish, but because of its use as a sewer "is very disagreeable."[52] Many of these farmers had come to specialize in dairy farming, marketing their milk and butter to neighboring manufacturing towns. The farmers needed the hay along the river to feed their cows over the winter and depended on clean water for their animals to drink. Unfortunately, the very towns that encouraged the farmers to specialize in dairying were also polluting the streams the farmers needed to maintain their cows.[53]

Before the 1860s, almost all tort cases involving water that made it to the Massachusetts Supreme Judicial Court were about quantity. Indeed, there were only three cases relating even vaguely to pollution before the supreme judicial court in this early period. After the 1870s, more and more of the cases involved quality issues. These were nuisance cases, or cases of downstream riparian users complaining that the water, after its use upriver, was compromised by that use.[54] Citizens not only complained about pollution in court cases; they also took their complaints to the newspapers.

When the Western Railroad moved its tracks and yard from the riverfront in Springfield, it offered the land to the city for a public park. A park along the river whose banks were described as having "absolute beauty" would have been a wonderful addition to Springfield in 1803, when the "elegance" of its shore was described, or even in 1819, when Benjamin Silliman described the shoreline's "extraodinary beauty."[55] But by the end of the nineteenth century, a gift of riverfront property was not welcomed by the citizens of the city. Recognizing that a park by the river's edge would be ugly and smell of pollution and waste, a letter writer to the city paper argued that parks were meant to be on high ground removed from the stench of the river, and that the city should refuse the offer.[56]

A Demand for Action

If the manifestations of pollution were obvious, the causes were diffuse. Water pollution came from many sources; several mills, workshops, and tenements. One irony of the fish cases was that when the state attempted to force dam owners to build fishways, the dam owners responded by attributing the declining fish populations not to the dams but rather to pollution caused by the mills and other industrial facilities that ran on dam-generated waterpower.

Traditionally, aggrieved individuals or communities sued the perpetrator of a nuisance in the court of common pleas or took direct action against the nuisance, as had the farmers who used crowbars to level offending dams.[57] When the offending agent was not a single individual or company, individualist solutions proved less effective. Whom did you sue when the cause of the nuisance was so varied? Even the traditional forms of resistance to nuisances were limited by court decisions that decided cases based on a shifting concept of "the greater good."[58]

To a great extent, pollution was the product of society's transformation from small-scale production to large-scale corporate industrial manufacturing. The change in production also produced intertwining interests, as the Massachusetts Supreme Judicial Court reminded tenement owner Royal Call. Call claimed that the steam engine powering a neighboring textile factory was a nuisance because its pollution drove away his tenants. Noting that Call's tenants worked in the factory in question, Judge Shaw ruled that if the court shut down the factory as a nuisance, Call would lose his tenants anyway.[59]

Multiple sources of pollution, complex patterns of interdependence, and conflicts of interest complicated the search for a solution to environmental degradation. As the Connecticut State Board of Health noted in 1880, "In the cases of streams, it is vastly more difficult [to protect purity] for a variety of reasons, not the least of which is that the manufacturing industries are affected, and moreover its relations to the general health is more extensive and varied."[60] The complication, though, did not eliminate the urgency of addressing the problem.

To overcome the conflicts of intertwined local interests, limits of traditional forms of redress, and growing complexity of pollution, New England health reformers, such as farmer and veteran abolitionist James Olcott or physician Henry Bowditch, pushed for a more aggressive, proactive state role in defending the rights of the citizens to "pure and uncontaminated air, and water, and soils."[61] Bowditch argued in 1872 that "streams are practically used as sewers . . . everywhere except in the face of the most stringent regulations and a police vigilance."[62] The Connecticut State Board of Health noted in 1880 that "it is certainly the duty of the government to protect the weak from oppression of the strong . . . and especially to protect that class called the poor. It is this last class that suffers most from . . . unwholesome surroundings and other unsanitary conditions which can only be controlled or suppressed by official effort. . . . It is only a question of time how long it will be before each state . . . must provide some official means to also protect the public at large."[63] Those concerned about the prevalence of disease believed that a cleanup needed to be orchestrated at a level able to oversee the larger setting. The orchestrating agency that had the vision of the larger setting was the state.[64]

Although the reformers believed that the state should take on the role of the protector of the public health, it was the reformers themselves who moved the state in that direction. That movement required political action and generated political conflict.

Bowditch and State Medicine

At the first September meeting following his appointment as chair, Bowditch rose to address the Massachusetts State Board of Health. In that speech, he outlined his idea for the board's function. He envisioned an activist board that would not "simply sit and gravely hear complaints of vile sights and fetid odors."[65] Instead, Bowditch argued, the board should go out "officially as an organ of state power" and take "vigorous action."[66]

Bowditch believed the newly created state board of health should be an active agent of state power practicing what he called "State Medicine." To Bowditch, state medicine meant acting "in light of the broadest philanthropy."[67] The board's job, he lectured to his fellow board members, would involve "thorough and scientific investigation of the hidden causes of diseases," and the board would work "to prevent the very origination of such disease."[68] Specifically, he noted, the board would investigate areas where disease seemed prevalent and root out the causes. It would hold public meetings relating to health. It would publish tracts with information about sanitation, health, and disease and report its actions and the state of health of the citizens in the commonwealth.[69] Bowditch wanted, as he wrote in a letter to his daughter, "nuisance makers in . . . the state [to] feel the power the legislature granted to us."[70]

Although there was European, particularly German and English, precedent for "state medicine," Bowditch knew all too well that in America, state medicine opened up a new and in many ways untested frontier in the alliance between the state and the medical community. Because it was new, Bowditch was also aware that the direction taken by this first board would have a tremendous influence on the future. "Our work is for the future as well as the present, and [in] all this many openings of our labors, we should . . . act wisely and for the ultimate good of the whole people."[71]

It was a bold agenda, but he felt the time was right for bold action. "Gentleman, let me say that I feel alike our grave responsibilities and the dignity conferred on each by his excellency the Governor."[72] If the board were successful in "strictly and impartially" pursuing its goal, Bowditch believed, "we should have a condition of public cleanliness and of public health, which would make Massachusetts a model for all other communities."[73] Bowditch's vision of state medicine combined the power of the state, the modality of modern science and its believed potential for "impartiality," and professional men committed to justice. For Bowditch, impartiality meant not lending their support to those who had privilege and power in the region. It did not mean avoiding conflict or being indifferent to the larger issue of justice. Bowditch had a modern vision of the state. He also had a radical notion of justice, infused by the experience of the Civil War and the enthusiasm of the early Reconstruction period.[74]

Bowditch was well aware that England had already made tentative steps in the direction of state medicine. For the other members of the board, Bowditch sketched out the brief history of state medicine and sanitary reform in England.[75] In the 1840s and 1850s, Sir Edwin Chadwick had argued that a healthy population needed a clean environment. Mobilizing concerned members of the elite and professionals for sanitary reform, Chadwick and his followers pushed for greater government action to clean up urban wastes, remove sewage pollution, and maintain clean homes and communities.

Within a year of the board's creation, Bowditch suggested to its members that he investigate public health work in England. By May of 1870, he was in London, linking up with reformers there.[76] Bowditch toured London's slums, where he saw the worst housing and squalor. The experience renewed Bowditch's faith in reform. "The authorities here undertake to provide shelter, and they make *vi et armis* men and women clean. In doing so, the health and morals of the community are elevated. Why may not by an extension of the principle, and on the grounds of self-defense too, the public authorities eventually build houses [for the poor]? ... Surely the houses in which the poor live and move and have their being ought in like manner be cared for by the public; if not for humanities sake, then for the sake of the public health."[77] Bowditch's vision of the role of the board of public health was expansive and, as is evident, included addressing not only the immediate cause of disease but also the larger environment in which the poor lived. If poor housing was "a plague spot, a center of the abominable miasma of contagion," then Bowditch believed that it was the state's responsibility to provide decent housing as an issue of public health.[78] But as Bowditch admitted, this idea was ahead of its time. "When shall we fairly grapple with this idea?" he wondered. "I have full faith that some time my dreams [of ending poverty] will be realized by others, even if I die, as doubtless I shall, without realizing one iota of the hopes for the future of humanity that press upon me when speaking of such life [poverty]."[79]

While in England, Bowditch also attended a conference of the National Association for the Promotion of Social Science in Newcastle, where he sat in on sessions on sewage and health.[80] He visited with reformers John Ruskin and Octavia Hill and the Peabody Settlement House and discussed sewage disposal with English experts.[81]

Bowditch returned from England energized about reform and more than ever convinced that the state should be the agent of reform and protection.[82] In that first address to the board of health, he argued that the board should guarantee that each citizen "not only have as long a life as nature would give him, but live as healthy a life as possible."[83] As the Massachusetts board noted in 1874, "It is manifest, then, that the first and largest interest of the State lies in ... the health of the people ... and the chief responsibility of the government is to protect it."[84]

The state also had the power to deal with the public good, "to reach out its paternal carefulness," the Connecticut board noted.[85] If health was a public good, and the citizens had a right to clean air, water, and soil, then the state had an obligation to protect that right.[86] State action was also needed because "local and private interests have often been so strong as to paralyze the action of the [local] health authorities."[87] As the New Hampshire Board of Health noted in its first report to the legislature, "No individual or association can undertake and carry alone such a labor [the protection of the environment] as well as the state."[88]

Believing that the poor had less power to control their environment, these early boards of health stressed the need for the state to act (paternalistically, to be sure). This was especially true as clean air and water were now compromised, and the poor did not have the financial resources to purchase what had once been free. These reformers believed that the poor, condemned to live in cheap housing near the factories where they worked and to drink water spoiled with wastes, bore the greatest costs of a compromised environment. In 1884, the New Hampshire Board of Health noted that "pure air and pure water are essential to the perfect development and good health of every individual. . . . The source of a water supply is generally a free one. . . . [But the poor] usually are obliged to rely on the most dangerous sources."[89]

In 1869, when Massachusetts created its board of health, Henry Bowditch noted that the legislature had "proposed a system ... capable of doing good."[90] In 1873, the Massachusetts board noted that "endowing [the board] with ample authority, [the state] has taken large and very wise steps. . . . All these show that the government recognizes its interest in, and responsibility for, the health and working power of the people and its determination to lend its authority for their promotion."[91] Connecticut's first board of health's secretary noted "a conviction on the part of those acquainted with the rapidly growing pollution of our streams that the time had arrived when the interference of the state jurisdiction was imperatively needed."[92] Increasingly, the Massachusetts Board of Health noted in 1873, people "even in some of the little cities of Massachusetts, that within a generation were open country villages," realized that the problems of pollution "need the vigorous arm of the law."[93] In 1882, the New

Hampshire Board of Health claimed that the legislature understood that "the interests of the people . . . are inseparably interwoven with the welfare of the state."[94]

In defending the new role of the state, health and antipollution reformers articulated a public right to a safe and clean environment. The Massachusetts board argued in 1869 that "all the citizens have an inherent right to the enjoyment of pure and uncontaminated air, water, and soil, and that this right should be regarded as belonging to the whole community; and that no one should be allowed to trespass upon it by his carelessness, or his avarice, or even by his ignorance."[95] In its first report, New Hampshire's board echoed the Massachusetts board: "We believe that every person has a legitimate right to nature's gifts, pure water, air, and soil,—a right belonging to every individual and every community, upon which no one should be allowed to trespass through carelessness, ignorance, or other cause."[96] Picking up on the longstanding practice of taking state action to protect industry, the board noted that public health needed as much state action and "consideration as given to the care of property or the fostering of productive industries."[97]

Rights to clean air, water, and soil demanded public action to clean up sewage and wastes from rivers and streams, and to limit smoke and air pollution, but for Henry Bowditch, there was more to healthy living than clean air and water. Drawing on his own need for a retreat to the seashore or a walk in nature in order to find peace and tranquility, he argued for the citizens' right to a reprieve from the pressures of work and urban congestion. "Every one ought . . . to leave his toil for a certain period each day and devote himself to the healthful recreation of walking . . . or fly to the woods or sea-shore." "A camp in the woods," Bowditch believed, should be available to each of the citizens for their renewal. He argued that children particularly "should be compelled to find some recreation in the open air."[98] To maintain a healthy community, Bowditch felt, "the state thus has an interest not only in the prosperity, but also in the health and strength and effective power of each one of its members."[99]

The reformers in New Hampshire argued that because the United States had a tradition of freedom that did not prevail in Europe, state intervention was all the more necessary. Without it, they believed, freedom could be used by business to take advantage of, rather than liberate, the individual.[100]

These reformers were well aware that the strong American reaction to the mercantile system of the pre-Revolutionary period had created an environment and ideology of limited governmental control. Americans had depended on common law and litigation to mediate conflicts within their contentious but open economic system. Yet as the New Hampshire board noted, because of the limitations and difficulties of common law, "people are always slow to avail themselves of such protection."[101] In the increasingly complex and interdependent industrial world New England was becoming, the state needed to act. "It is [our] function," the New Hampshire board argued, "to furnish the weapons for this fight [against air and water pollution]."[102] The Connecticut board concurred, arguing that "there seems to be a need of . . . legislation upon this subject, especially when the water is so contaminated by manufacturing wash and sewage to be detrimental to health."[103]

Water Pollution and Reform

Identifying water pollution as the most critical problem before the commonwealth, the Massachusetts board in 1871 conducted an extensive study of the pollution in Millers River. The next year, the board looked at the impact of pollution on the state's major streams.[104] In 1872, the legislature instructed the board to report on sewage, stream pollution, and the water supply of towns. Hoping to restore the "economy of nature," the legislature also asked the board to investigate whether sewage could be converted into fertilizer. The state was concerned about pollution that "reaches to the very foundation of national health and prosperity" and "presses with great force upon the people of the state at the present time. Our manufacturing and commercial industry are growing with unexampled rapidity. . . . Some of the brooks which were but recently pure and undefiled are now polluted so that neither man nor beast will freely drink of them; and this change is insidiously taking place from year to year."[105]

Following these early reports on the critical state of water supplies in Massachusetts, the legislature in 1875 directed the board to focus attention on river and stream pollution. The board appointed James Kirkwood, a Brooklyn, New York, engineer; William Ripley Nichols (who had done the earlier 1872 survey with George Derby); and Frederick Winsor to survey the state's waters, evaluate the pollution risks, and pinpoint pollution problems. In particular, the board asked the investigators to look at industrial pollutants dumped into the water systems. The report the board sent to the legislature in 1876 was more than four hundred pages long. It contained a detailed analysis of water quality, highlighting the impact of industrial wastes on the investigated rivers. It also included a history of pollution reform in England.[106]

Investigating pollution inevitably led to assigning responsibility. Pollution may have had diffuse sources, but there were still identifiable polluters. Ultimately, if the reformers wanted to successfully maintain "clean air, water, and soil," they had to confront the polluters. First, however, the polluters had to be identified. The investigators for the board of health discovered two major sources of pollution: sewage wastes from homes and municipal sewer lines, and industrial wastes dumped by manufacturers.

In an extensive 1877 study of the sources of pollution of the state's streams, board member Charles Folsom attempted to locate specific problems. In his final report, "The Pollution of Streams, a Report of Sewerage," Folson investigated the pollution problems of each town in the commonwealth to determine possible solutions.[107] In 1878, the board took Folson's information, along with its study of state actions in European countries, to the state legislature. In an attempt to strike a moderate tone, the board noted that the means of cleaning up water pollution were "almost unknown in this country," and that "scarcely a beginning has been made here." They also noted that "it would be impractical to wholly stop at once the pollution of streams by means of methods untried by us." That said, the board did argue that "a first step [should] be taken by the State, and that even more general restrictive measures may be applied a few years hence."[108]

The board then pushed for legislative action to begin cleaning the state's rivers and streams. Under the proposed law, neither individuals, corporations, or cities "shall discharge, nor cause to be discharged . . . into any stream or public pond . . . any solid refuse . . . or any polluting substance so as either singly or in combination with other similar acts . . . [to] pollute its waters." The act would prohibit a stream from being "converted into a sewer in any city or town." The act outlawed the discharge of sewage or refuse containing human excrement and also required that discharge from manufacturing or other establishments be "cleaned and purified." The act allowed present polluters a "reasonable length of time to comply with the provisions of . . . [the] act." It would also create a rivers pollution commission to monitor water pollution, approve or disapprove sewage plans, issue cease-and-desist orders in cases of public nuisances, and provide permission for action.[109]

Despite resistance from manufacturers, the legislature followed the board's recommendations and passed the Act Relative to the Pollution of Rivers, Streams, and Ponds (Acts and Resolves 1878 c 183). Although the act gave the board sweeping power, it also reflected the interests of the largest of the manufacturing corporations. It exempted the two rivers of the state that had the greatest concentration of industrial use—the Connecticut and the Merrimack (including the Concord River, within the city limits of Lowell)—and by either prescription or legislative grant allowed continued polluting by some corporations.[110]

Even with these compromises to the industrial water users, by identifying corporations as polluters, the board raised the issue of public good over private good.[111] Reformers were reluctant to argue openly that corporations polluting waterways should be shut down. But they did suggest that corporate and public interests competed and "with our present imperfect knowledge of the best methods of utilization [of wastes] may even stand for the time in direct opposition to each other."[112]

And in opposition, they did stand. No sooner had the water pollution bill passed the state legislature than manufacturers began to lobby not only against the bill, but against the board of health itself. Henry Ingersoll Bowditch envisioned an aggressive state board, practicing "state medicine" for the public good, and protecting the rights of the citizens to pure air, water, and soil. In practicing state medicine, however, Bowditch aroused the antagonism of the manufacturing interests, who identified public good with their own well-being and success. Unlike Theodore Lyman, who saw his role as mediating between the manufacturing interests and the needs of the fisheries, Bowditch considered himself a crusader for the weak, the poor, and the defenseless. To Lyman, science and technology transcended competing interests. But Bowditch believed that science should inform the public in its struggle for the common good. Lyman may not have gotten fish up the region's rivers, but neither did he open up a hornet's nest of political conflict and intrigue. That is exactly what Bowditch did.

7

Cooperation, Conflict, and Reaction

When James Olcott spoke before Connecticut farmers for "anti-stream pollution," he urged the public to mobilize to stop water pollution by "ignorant or reckless capitalists."[1] In identifying the "ignorant and reckless capitalists," Olcott focused the attention of the farmers on industrial waste and the role of manufacturers in their search for profits in causing pollution. Although manufacturers and the courts argued that industrialization brought wealth and prosperity to New England and hence was a general good, Olcott challenged this idea. He saw the issue as a conflict between industrialization and its costs on the one hand and the public good on the other.

Concern over industrial pollution and the potential conflict between it and public health had already arisen in Massachusetts. Although the Massachusetts State Board of Health realized that the interests of the "capitalists" and those of the public health officials might be in conflict, in 1872 it hoped that with improved knowledge, "a way will be eventually found to joining them into harmonious relations," much as Lyman believed science and technology would resolve the conflict between fishers and mill owners.[2] The board's interest in "harmonious relations" also reflected a realization that at least for the last several years, the courts had seen pollution as an inevitable consequence of civilization and had been favorable toward industrialists, especially if no obvious alternative to dumping pollution existed.

In 1866, William Merrifield sued Nathan Lombard because Lombard had dumped "Vitriol and other noxious substances" into the stream above Merrifield's factory, "corrupting" the water so badly that it destroyed his boiler. Chief Justice Bigelow ruled that Lombard had invaded Merrifield's rights. "Each riparian owner," the judge wrote, "has the right to use the water for any reasonable and proper purpose.... An injury to the purity or quality of the water to the detriment of the other riparian owners, constitutes in legal effect, a wrong."[3] In 1872, Merrifield again went to court, claiming the City of Worcester regularly dumped sewage into Mill Brook, by which the waters became greatly corrupted and unfit to use." Supreme Judicial Court Judge Wells ruled in this case that although the city was "polluting the water," the question was "upon what grounds and to what extent is a city responsible in damages for such effects." Wells claimed that riparian right was "not an absolute right, but a natural one, qualified and limited . . . by the existence of like rights in others. The judge ruled that Merrifield's right to water "in its pure state . . . must yield to the equal right in those who happen to be above him. Their use of the stream . . . will tend to render the water more or less impure." Development, the judge noted, "unavoidably cause[d] impurities to be carried into the stream . . . , but so far as that condition results only from reasonable use of the stream in accordance with common right the lower riparian proprietor has no remedy. When the population becomes dense and towns . . . gather, . . . the stream naturally and necessarily suffers still greater deterioration." The judge then ruled that to prove his case, Merrifield would have to demonstrate that Worcester failed to maintain the city's sewers. It was acceptable to pollute a stream as long as you were doing it according to the existing science and technology.[4]

Merger of the Boards

Olcott, who gave his talk fourteen years after the Massachusetts board optimistically hoped that with improved knowledge, "harmonious relations" might exist between the manufacturers and antipollution reformers, was less sanguine of that possibility. Olcott saw that the powerful interests, particularly the industrialists, opposed water cleanup. He believed that the reformers had to agitate to mobilize the common people against the polluters, the same way the abolitionists mobilized against the slave powers.[5] It may have been the example of what happened in Massachusetts that made Olcott pessimistic about the possibility of cooperation between the industrialists and the water purity reformers.

In 1878, shortly after the Massachusetts board passed the Act Relative to the Pollution of Rivers, Streams, and Ponds, agitation began over the state's "spending wastefully the people's money" on its numerous commissions, especially the board of health.[6] Facing a stiff election challenge from General Benjamin Butler, Governor Thomas Talbot decided to show his concern for economy in government by merging the board of health with the board of lunacy and the board of charity.[7]

Under the umbrella of the new merged board, the interests of the pollution reformers were lost. Bowditch, the former chair of the board of health, was now only one of many members of the new expanded board.

Charles Francis Donnelly, formerly the head of the board of charities, emerged as the chairman of the new board, and Donnelly had personal and ideological reasons for not pursuing an aggressive policy toward industrial polluters.[8]

Initially, Bowditch complained that the issues of public health, particularly pollution, were being ignored by the enlarged board of health, lunacy, and charity. In a letter to his son, Bowditch complained that "our state board is . . . united to the Board of Charities. Our meetings are interminable upon questions relating to everything but sanitation. I get mad every time I go to a meeting, and have about made up my mind to resign."[9] Before resigning, Bowditch lobbied unsuccessfully to have the board of health separated off again as an independent board. "Our efforts were fruitless. I, finally, as a solemn protest against the absurd and fatal combination, resigned after . . . fruitless effort to persuade a change."[10]

Before resigning, Bowditch also struggled to get pollution issues back on the agenda and pushed the new board to enforce the antipollution act aggressively. Despite his efforts, the new board failed to act against polluters. Bowditch came to realize that the problem with the new merged board was not simply inefficiency. He believed that the cost-savings argument for the merger of the boards was only a smokescreen. The real reason, in his mind, for the merger was the hostility of manufacturers to the aggressive antipollution positions of the board of health under Bowditch. The new merged board "fell into the hands of self seeking capitalists who were afraid of the millstreams being cleaned from the pollutions poured into them by the mills owned by evil capitalists."[11]

The person Bowditch felt was most responsible for the failure of the merged board to take action against polluters was its new chairman.[12] "It is currently understood that a new chairman, previously a high official in the state [board of charities and] chief of a large factory situated on the banks of one of our principle rivers and into which he poured his polluting refuse, said sneeringly, 'do not [think] that I allow myself to be chosen chairman of this Board with the intention of permitting anything to be done by it on the pollution of streams.'"[13]

The new chairman, Charles Francis Donnelly, was born in Ireland, October 14, 1836. One year later, he came with his family to New Brunswick, Canada. Donnelly and his family moved to Providence, Rhode Island, when he was twelve years old. Donnelly graduated from Harvard in 1859 and began a career in law, first in New York and then in Boston.[14] Once in Boston, Donnelly became involved in Catholic Church issues, and he was the legal advisor to Bishop Williams of Boston.[15] Donnelly began working in state charity when Bishop Williams asked him to protect Catholic orphans from being placed in non-Catholic homes. Besides his legal work for the church, Donnelly served as a lawyer for several manufacturers.

He also had family connections to textile manufacturers.[16] Donnelly shared the belief of many that manufacturers brought prosperity to the region, and that the state should not be involved in private affairs, whether those were the affairs of manufacturers or of individuals. Donnelly, a conservative Democrat, opposed the liberal-activist Democrat Benjamin Butler.[17]

Following Donnelly's appointment to the chair of the merged board and Bowditch's resignation, the reformers fought back. The reformers were well positioned in the state. Many came from leading families and, like Bowditch, had personal ties to leading politicians. They socialized in the same circle as those who held influence within the state. But so did the manufacturers. In opposing the industrialists, the reformers confronted not only their political opponents but also their neighbors and relatives. This made mobilizing support more difficult when they tried to use their position as leaders of the community. Bowditch knew many of the state's governors personally. Unfortunately for him, those he attacked as "evil capitalists" also had social standing in the community, a friend in the governor's chair, and influence in the legislature. Failing through personal contacts to persuade the governor to act, the reformers took their case to the public at large.

Public health activists' hostility to Donnelly, as well as a conflict between the conservative Donnelly and the new governor, Butler, led Butler to remove Donnelly as chair when his term expired in 1883. To appease the reformers, Butler appointed the public health activist Henry Walcott of the American Public Health Association to the health committee of the merged board.[18] Butler was no novice in the area of public health. He had already established a reputation for a concern for the health of the general community during his occupation of New Orleans. He believed that the government had a responsibility in that area.[19] In Henry Walcott, Butler found a person who shared his concern for public health.

Henry Walcott: Patrician Reformer

In many ways, Henry P. Walcott was the perfect candidate for the board. Unlike Bowditch, Walcott did not have a reputation as a troublemaker, and unlike many of the region's reformers, Walcott was friendly with Governor Butler.[20] Yet Walcott was a distinguished doctor with a solid record of public service and came from a leading Massachusetts family. After graduating from Harvard in 1858, in the class before Donnelly's, Walcott continued on to Harvard's medical school, eventually transferring to Bowdoin, where he completed his medical training in 1861. After two years in Europe furthering his studies, he returned to Massachusetts to set up a practice in Cambridge. There, he spent his time practicing medicine, serving the community and Harvard University, and socializing with the Cambridge intelligentsia.[21] Walcott's interest in and public support of Shattuck's 1850 study of the state of health in Massachusetts and Walcott's belief in public service led to his appointment in 1882 to the merged board of charity, lunacy, and health.[22] When Donnelly's term expired as chair, Butler chose Walcott to assume the position.[23] The same year, Walcott was elected president of the American Public Health Association.

While the health committee of the new board, and particularly Walcott, pressed for action on pollution, manufacturers accused the health committee of hostility to their interests and the economic development of the state.[24] In 1882, one manufacturer argued before the state legislature that pollution enforcement would compel the manufacturing interests to move out of the state, leaving behind unemployed "villagers which depend upon the mills for their prosperity."[25] Although manufacturers had successfully weakened the 1878 antipollution law, they still saw the law as a threat to their customary habit of dumping wastes into the nearest waterway. Although the higher courts were generally sympathetic to polluters, the 1878 law did raise a serious question about the ability of industry to claim the right to pollute under prescription. In 1882, in *Brookline v. Mackintosh*, the Massachusetts Supreme Judicial Court, although ruling in favor of Mackintosh, who was dumping arsenic and other wastes into the river, commented in its ruling that an industry, however "useful and necessary" and even with a longstanding practice of polluting the water, could be prevented from doing so if the state felt the polluting was "incompatible with the health, safety and welfare of the community." Moreover, the court noted that since the 1878 law, an industry could not gain the prescriptive right to pollute, since there was no right of prescription gained by engaging in behavior that violated statutory law.[26]

With this decision, the court seemed to open the door for legislation prohibiting or restricting pollution. It was a position the reformers had long advocated. The manufacturers did not miss the implications of this ruling either. For both sides, the possibility of "harmonious relations" seemed remote. Instead, both reformers and manufacturers mobilized to promote their own interests. In such a context and even without Donnelly as chair, the merged board stalemated in a conflict that pitted the medical and reform community against the manufacturers. The *Boston Medical and Surgical Journal* editorialized that the manufacturing interests of the state were corrupting politics to get away with polluting public water, leaving the poor to "drink bad rum." The *Journal* complained that manufacturers prevented public action through their "control over politics."[27]

The Antireformers Fight Back

In 1883, Butler lost his reelection bid, and Republican George Robinson from Chicopee took office.[28] Manufacturers made their hostility to Walcott clear.[29] Anxious not to alienate the manufacturing interests, Robinson reappointed Donnelly as chair. Robinson refused even to reappoint Walcott to the board.[30]

Bowditch and the reformers were furious. Bowditch went to the governor to beg him not to capitulate to the industrialists in removing Walcott, "the only person on the board working honestly for sanitary ideas."[31] Robinson refused to reconsider. He did ask Bowditch for a recommendation for a replacement for Walcott. Bowditch refused. The governor ended up appointing a replacement who Bowditch felt "had never manifested any interest in sanitary disease."[32]

Meanwhile, Walcott, with the support of the antipollution reformers, fought his removal himself.[33] He denounced the corrupt influence of the manufacturers on the board and called for a new and even more rigorous antipollution act. Walcott had a reputation within medical and scientific circles as an honest scientist who had integrity, and he used that reputation to help mobilize support. Although the state's manufacturing interests carried significant clout and were heavily represented among Boston's elite, Walcott also had influential friends. Like Lyman and other intellectual Boston Brahmins, Walcott was a member of the Thursday Evening Club. He was president of the Massachusetts Medical Society, president of the Massachusetts Horticultural Society, and a member of the state forestry association, the Massachusetts Historical Society, and the Union Club of Boston. The *Boston Globe* characterized Walcott's cause as a campaign for "purity over impurity," as public support for the scientist increased.[34] An issue that the governor had hoped to contain soon spilled into the larger public arena, as interest in the conflict grew.[35] The Massachusetts Medical Society appointed a committee to gather petitions from around the state to have the board of health reestablished.[36]

Realizing the growing public interest in the matter, the *Boston Herald* assigned a reporter to do a series of stories on the conflict. In response to inquiries by the reporter, Donnelly articulated a vision of public health and the state diametrically opposed to Bowditch's view. Bowditch believed in the agency of the state to meliorate the problems of pollution and disease. Donnelly, on the other hand, saw health as an individual matter. For Donnelly, the state had no business addressing issues of health. "The Physician examines and experiments that is all. It is contrary to American ideas for the state to take care of the health of the people. By doing so self reliance is taken away. The average citizen needs no state board."[37]

The petition campaign and public support for the "purity" campaign, as well as hostility to Donnelly brought on in part by his intemperate comments to the *Boston Herald*, finally led the legislature to recommend to Governor Robinson that he recreate the board of health as a separate board with enlarged powers and duties. Robinson relented and created an independent board of health with "more responsibility for the purity of our rivers."[38] He also appointed Walcott to head the board.

With public opinion aroused and the reform community mobilized, the state passed a new and stronger water purity act in 1886. This act was specifically designed to protect the purity of water in streams and ponds used for drinking. It gave the board of health authority for the maintenance of water quality for all the inland waters of the state.[39] Two other reform doctors were appointed along with Walcott to the new board.[40] Although the reformers won a victory in having the board reestablished with Walcott as chair, the governor, in order to placate the manufacturing interests, also appointed to the board two manufacturers and Hiram Mills, the chief engineer for the powerful Essex Company of Lawrence.[41]

Charles Donnelly's ideas about the role of the state in health care found their roots in a vision of individualism that was deeply embedded in the American culture. It was a vision that, when linked to ideas of progress and development, had a significant impact on how one understood the environmental change occurring throughout the northeastern section of the nation.

Even as Bowditch and the board of health argued for a more aggressive state role, they realized that American beliefs in and about individualism might undermine the ability of the state to act for the public good. The board of health asked rhetorically, "What and where, then is the responsibility of the state?"[42] Bowditch reminded Massachusetts residents of the rapid change that had occurred around them. In 1800, only 6.8 percent of the population lived in towns of more than ten thousand. In 1840, 22 percent lived in cities over ten thousand, and by 1870, almost half the state's population lived in such communities. "There is a great tendency, promoted by civilization and increasing wealth and industry," he argued, "to gather people in compact villages for manufacturing purposes." And, Bowditch reminded his fellow citizens, the "death rate keeps almost constant pace with increasing density." Faced with this crisis, the state had to act. "As a drowning man, or child falling into the fire, demands help, prompt and energetic in proportion to the imminence and degree of the peril, so some sanitary dangers demand the immediate and efficient interference of the law."[43] Although he admitted that industrialization brought wealth, Bowditch believed that human life had precedence over monetary gain. "The law now authorizes and commands the boards of health to make these reforms, at any pecuniary cost, for human life and strength are not to be weighed in the same scale with money."[44] Bowditch argued that society needed to be protected against the irresponsibility of human greed, exemplified by the industrialists' seeming indifference to public health.

If for Bowditch, Olcott, Walcott, or even Marsh the increased urbanization and industrialization of the region called for a more interventionist state, Donnelly adamantly disagreed. Perhaps Donnelly's suspicion of state intervention had its roots in his suspicion of the motives of New England's Protestant majority, as well as his economic views. When the Massachusetts legislature considered a bill to inspect schools, Donnelly suspected that the motive for this bill was to interfere with Catholic instruction in parochial schools, not to improve education.[45] As a lawyer for a minority, Irish Catholic population, Donnelly was concerned that the majority Protestant population would use the power of the state to interfere with the private religious affairs of the individual. Donnelly's suspicion of state action also fit with his personal convictions.[46] But Donnelly's concern over state intervention involved him in support of a particular vision of "American ideas." According to this vision, the state would not interfere with a person's religious beliefs or in the affairs of private corporations. Nineteenth-century American individualism encouraged the kind of antistatism of Charles Donnelly. In a land filled with opportunity and resources, Donnelly believed the state had a limited role. Individuals had responsibility to take care of themselves.

Charles Donnelly saw the state as an agent proselytizing to Catholic children and harassing Irish immigrants. Like Donnelly's, Thoreau's antistatism came from a deep suspicion of the motives of those involved in government actions. Yet Thoreau also cherished nature's abundance and suggested, in a rare optimistic mood, that communities should take control of their forests and maintain reserves.[47] Although Thoreau would probably look with cynicism at the efforts of Henry Ingersoll Bowditch, he certainly would not have joined Donnelly in defending the rights of manufacturers to pollute. Thoreau's critique was essentially antimodernist. Donnelly, on the other hand, was a developmentalist who felt the state should support but not restrain economic expansion.

In this, Donnelly had the support of the manufacturers. Manufacturers argued that pollution (and in the case of the dams, the decline of fish) was not their responsibility but a general consequence of growth and development. In a sense, they were correct in their analysis. The manufacturers' argument attempted to associate progress and development with inevitable environmental decline. If industrialization inevitably meant more pollution, then either one had to accept environmental degradation, or, it seemed, one had to reject progress and development themselves.

Like the manufacturers who testified before the Massachusetts legislature that cleaning up their pollution would force them to move out of the state and leave behind unemployed villagers, those opposed to pollution reform wanted to frame the issue as between industrialization on the one side, and clean air, water, and soil and abundant resources on the other. This way of defining the issue divided the conservative reformers like Lyman from those who opposed any attempt at state action. The manufacturers rejected the notion that one could have industrial development and at the same time husband resources and protect clean air, water, and soil. But mostly they believed that any attempt of the state to interfere with their workings, which they defined as the natural workings of the economy, would hurt them.

These industrialists who opposed state action for the environment and their supporters were not laissez faire theorists. They believed in a state that maintained a strong tariff. Their industries had grown up around protection and government support. Let us not forget that it was a Massachusetts state lottery bill that rescued the original dam project at Hadley Falls. What these manufacturers believed in was whatever was in the interests of business. They understood their interests. They supported state action that furthered their business interests. They opposed it when they believed it would hurt their interests.

The manufacturers argued that their business interests were in the public good. They claimed that their mills and dams, even with the accompanying pollution and resource depletion, provided jobs, opportunity, and growth. For those who found jobs, built homes, and sold building lots, the mills and factories and towns and cities that grew up around them provided beneficial results. The industrialists saw in the workings of industrial capitalism not only their own financial success but also progress and prosperity for the whole community.[48]

Henry Ingersoll Bowditch also understood interests and progress. But Bowditch posed a different interest, a different understanding of progress, and a different good. He claimed that the good of mills, factories, and cities could not be separated from the interests of the citizens in good health and a clean environment.[49] Indeed, Bowditch claimed that the health and environment of the citizens was a greater good and needed to be privileged over the interests of the manufacturers. Progress for Bowditch should be measured in life expectancy and happiness. "The progress of civilization is best manifested in the progress of vitality."[50]

For Bowditch, the state was a means of achieving the greater good. It was also a means of opposing the interests of the manufacturers. Bowditch was not naive about the ability of the state to act for the public good. He had seen the state act to defend slavery. He saw the state capitulate to the industrialists when it merged the board of health with the board of charity and lunacy, and when it removed Walcott from the board. Bowditch also had seen the effect of mobilizing public support, and the ability of the board to exercise state power for public good.[51]

Bowditch had a radical vision of reform that included interests in conflict. This social conflict view he shared with the manufacturers. Unlike the manufacturers, though, he supported the interests of common citizens—their rights to clean air, water, and soil, and even the right of scenic retreats—against the interests of the manufacturers to increase profits.

Unlike Bowditch, Theodore Lyman believed that conflicting interests could be harmonized. He felt that with the proper application of science and technology, industry could have its profits, and the public could have cleaner water and more fish.[52] Lyman put his faith in the neutral objective scientist as the proper agent for bringing together the interests of business and the interests of the general public. Although he recognized that he and his family had always been identified with manufacturing, he held that his scientific training would transcend any personal interest he might have.

Science was as important to Bowditch as it was to Lyman. Both men were trained empiricists. Like that of Shattuck, Walcott, and Davis, Bowditch's and Lyman's science was one of observation and classification. They understood it in holistic, ecological terms. Although they did experiments, such as Lyman's test of pollution on fish, they were more comfortable with general observations.

Although both men shared a common commitment to science, they saw it functioning differently. Lyman saw science as a means of transcending interests, and as a means of maintaining neutrality. For Bowditch, it was a tool to mobilize popular support for the common good. As he told his fellow health board members at their first meeting, the board should act as "a special function of state authority [public health] which, until these latter days of scientific investigations, has been left almost wholly unperformed."[53] Science, as improvement in agriculture demonstrated, once mustered for the common good by the state could be of aid to society. "The government enlists the cooperation of men of learning, scholars in all collateral sciences, philosophers, naturalists, botanists, chemists, mineralogists, geologist, ornithologists, and entomologists. . . . Thus the state has, in manifold ways, obtained the aid of the science of her scholars. . . . In this work the legislature and people have gone hand in hand, mind with mind, heart with heart to effect their common purpose."[54]

But Lyman and Bowditch were not the only ones looking to science as a means to an end. The industrialists whose dams blocked fish migrations and whose factories dumped "noxious wastes" into the air, rivers, and streams also had access to scientific knowledge.[55] Science and technology for these manufacturers would also help them in their never ending quest to control nature for greater profit.[56] It facilitated the building of their dams and the smooth running of their machinery. It also could be used to justify their actions and mute popular opposition. When Elizabeth and Hannah Emerson became sick because of gas fumes escaping into their home from the Lowell Gas and Light Company pipelines, they sued the company. Unfortunately for the Emersons, when they had two doctors testify that the gas was making them sick, the court ruled that the doctors were not experts on the impact of gas on health and threw out their testimony. The gas company then had an "expert" who was accepted by the court testify that gas fumes made a neighborhood healthier by reducing cholera and yellow fever.[57]

Bowditch, Lyman, and Donnelly represented conflicting visions of the role of the state and of science and technology. All three of those visions left an imprint on the national reform movement that emerged at the end of the nineteenth and the beginning of the twentieth century. As New Englanders confronted an environment ever more compromised by industrialization and urbanization, they attempted to fashion a means of dealing with the environmental change. What they fashioned included elements of all the visions about the role of the state, the environment, and science and technology that they had experimented with in the immediate postwar period. Because these visions were themselves conflicted, the solutions were also contradictory.

It has been argued that the post–Civil War reformers saw the state as a more activist agent because of their experience during the war.[58] Postwar reformers did have a far more aggressive view of the state, although the degree of aggressiveness differed tremendously, as we can see in the examples of Bowditch and Lyman. How important the war experience was in causing this change is unclear.[59] For people like Bowditch, Davis, and Olcott, the abolitionist movement was a life-shaping event. Yet Bowditch's views of the role of the state were also formed in France, during his stay there. They were affected, as well, by what was happening in England, for these reformers were internationalists and functioned in a truly transatlantic intellectual community. The Massachusetts State Board of Health's 1873 report not only listed the laws concerning public health enacted in England from 1839 to 1862 but also commented extensively on Chadwick's 1842 report on the sanitary conditions of the English laboring classes. It summarized the 1844 report of the health of English towns, and the various reports of the British Board of Health. References to English reform activities and research regularly appeared in the writings and published reports of reformers on this side of the Atlantic.[60]

The reformers were also Americans and influenced by the culture and ideology of this country. The various reform positions and ideas that emerged in the postwar period were informed by the American ideology of individualism and self-reliance that Donnelly espoused. They were influenced by activities, ideas, and programs in France, Germany, and particularly England. The changing environmental conditions at home also influenced them.

8

Industrial Waste, Germs, and Pollution

The Battle over Pollution

In 1905, the state board of health for Connecticut looked back over the last half century and noted the tremendous change that had occurred. In the first half of the nineteenth century, "all the towns and cities in Connecticut were very rural in character, and nowhere were populations so dense from overcrowding as to affect the public health. Hence there was no conspicuous disparity in the salubrity of different towns."[1] As Connecticut industrialized and urbanized, disparity in the salubrity of different parts of the state increased. It became "an accepted fact, sustained by careful observation, that the death-rate was always higher in cities than in the country."[2] Although the pure past to which the Connecticut State Board of Health alluded may not have been as pure and healthful as it assumed, nonetheless, the board was correct in noting the increase in mortality in the industrial towns and cities that grew up over the century. Growing awareness of the "effect of environment and employment upon the prevalence of . . . disease" created momentum for public action.[3]

The vision of an activist state promoting public health and protecting the citizens, particularly the "weak" and "poor," from the vagaries of the market—whether those were represented by "foul" water or depleted resources—increasingly found support among other reformers. The urban industrial setting that made Connecticut's cities so unhealthy also generated concerns over working children, long working hours for women in the paid labor force, industrial diseases, and overcrowded tenements.[4] Like the antipollution reformers, those who were concerned over these conditions increasingly looked to the state to legislate remedies. Laws that limited women's working hours and child labor and that controlled the conditions of tenements found favorable hearings among legislatures attuned to an electorate demanding reform of the conditions they found in their daily lives. Environmental reformers—both public health activists and supporters of protection for fish—were important voices in this rising chorus that favored a more active state. The momentum for public action began in Massachusetts, the most industrialized and urbanized New England state, and spread to the other states of the region and ultimately to the entire nation.[5]

While the Massachusetts Board of Health struggled to maintain its independence, the other New England states took up the challenge of state medicine. Faced with similar problems, these states soon created their own public health boards. New Hampshire, Connecticut, and Rhode Island had growing industrial cities that manifested all the problems documented by the Massachusetts Board of Health. Doctors and reformers in those states read the reports coming from the Massachusetts board and lobbied their own legislatures for their own boards of health.

In 1877, eight years after the creation of the Massachusetts board, the Connecticut legislature created the Connecticut State Board of Health. New Hampshire established its board of health in 1881. In 1886, the Vermont legislature created that state's board, which "should take cognizance of the life and health among the inhabitants of the state, investigate diseases and epidemics and the causes of death and the effect of employment on health." It also charged the board "to advise with municipal officers with regard to . . . the drainage and sewerage of towns and cities."[6]

The other New England boards followed Massachusetts's lead and began investigations of polluted water and air. They also lobbied state legislatures for increased powers of enforcement, and they pushed the idea that public health was a right that needed to be protected by an activist state.[7] They believed that each citizen had a right to clean air and clean water. But, as in Massachusetts, the boards faced stiff resistance. Corporations resisted pressure to stop dumping wastes into local streams and smoke and noxious fumes into the air, and towns resisted pressure to stop dumping their wastes into the nearest running water. As long as the corporations and towns could successfully argue in court that there was no alternative to dumping their wastes into local streams, public health officials had trouble stopping pollution, even where they had widespread popular support.

Manufacturers argued, against the public health advocates, that they had no choice but to expel waste into existing streams. It was an argument that did not convince the reformers. Despite the manufacturers' resistance, the reformers continued to push for cleaner water. After Governor Robinson recreated the board of health in Massachusetts as a separate entity, reformers were convinced that under Walcott, the board would again move against polluters.[8] Indeed, one of the first acts of the new board was to send out a letter to local governments noting that health laws would be rigorously enforced to "prevent their infringement by individuals, corporations or municipalities either from ignorance, carelessness or selfishness."[9] The re-creation of the board of health and the new pollution bill of 1886 gave the reformers a sense of vindication.

Although the antipollution bill may have been weakened in the legislature, the principle of the bill was established. The board members believed that it was only a matter of time before tighter legislation would be passed, and industrial as well as human waste would be removed or kept out of the region's streams.[10] Industrialists shared that feeling. As a result, they lobbied fiercely against the bill and the board. To stop dumping industrial pollutants into the rivers would increase costs. Industry had no interest in adding to its costs. For reformers like Hiram Mills, an engineer employed by manufacturers, the answer lay not in confrontation but in finding a means of protecting health without antagonizing the manufacturing interests. Although he did not come from a Brahmin family as Lyman did, Mills was a positivist who shared Lyman's faith in the ability of science and technology to provide progress and transcend conflicts of interest.

Hiram Mills: Technology and Reform

Hiram Mills was born in Bangor, Maine, in 1836 and after attending public schools went off to Renselaer Polytechnical Institute in Troy, New York, where he earned his chemical engineering degree in 1856. After graduation, Mills went to Brooklyn, New York, to work under James Kirkwood (the coauthor of the 1876 report on water pollution in Massachusetts). Later, Mills had worked under James Francis at Lowell until the Essex (textile) Company of Lawrence took him on as their chief engineer in 1869. Mills believed in waterpower and the advantages of technology in bringing wealth and prosperity to the region.[11] As a member of the board of health, he felt that he could bring his technological knowledge to the problems of disease and ill health.

Mills's faith paid off with the development of a new understanding of disease causation that led to the deemphasis of concern over industrial wastes, and a new way of dealing with the problems of typhoid, dysentery, and cholera. On the board, Hiram Mills became the head of the committee on water supply and sewage, a position he would hold until 1914. He used his position as head of the committee to focus attention on the problems of sewage treatment and the public health risk of typhoid and other sewage-borne diseases.[12] Lawrence, like Holyoke, Chicopee, and Lowell, suffered from problems of cholera, diarrhea, dysentery, and typhoid. Hiram Mills was convinced that the source of these problems lay in the drinking water of the city.

Mills was successful in dealing with the problems of pollution because he was able to develop a means that provided a partial solution to the problem of sewage waste. He was also successful because his concern directed attention away from industrial pollution and toward sewage pollution. In doing so, he avoided the intractable conflict between the interests of the manufacturers and the interests of public health. Mills's accomplishment rested upon the newly emerging acceptance of the germ theory of disease causation. It also depended on a new generation of scientific and technological specialists.

The Germ Theory

Although activists in the public health field believed they could see and smell the source of disease in polluted air and water, work in Europe in the 1870s and 1880s by the Germans J. Henle and Robert Koch, the Englishman Joseph Lister, and the Frenchman Louis Pasteur indicated that invisible germs were the source of disease.[13] By the late 1870s, the European theories of disease causation gained some attention on this side of the Atlantic, although many remained skeptical. In 1877, the Massachusetts State Board of Health admitted that there might be validity in the germ theory, yet it continued to argue that "certain diseases, especially cholera and typhoid fever may arise de novo under certain conditions of filth."[14] Public health officials in Connecticut questioned the validity of the germ theory into the early 1880s.[15] They were willing to entertain it, but only as a contributing factor because "the germ theory is not yet proven."[16] In 1883, the New Hampshire State Board of Health reported 204 deaths in Manchester of cholera, typhoid, and scarlet fever due to "miasmatic conditions."[17] As late as 1887, the New Hampshire board stated that "the fact [is] that this . . . germ theory of disease has assumed so much importance during the last few years," yet nonetheless the board still believed that "a division of scientific opinion as to what renders water dangerous to health . . . [exists]," and that "disease may originate like typhoid fever from filth de novo."[18] In its first report, the board attempted to accommodate the two conflicting theories, while at the same time giving primacy to what it believed was the more reasonable theory: "Whatever theory may be the correct one, the broad fact that filth is the foe to health cannot be controverted. It matters not to what extent the controversy may be carried or how many speculative theories are presented."[19]

Yet it did matter which theory prevailed. The growing acceptance of the germ theory changed how "filth" was seen and how it was assumed to act as a foe to health. It also changed how one responded to that filth, and indeed, how one defined it. By the end of the 1880s, the germ theory had won over most public health officials in the battle of disease causation.[20] In Connecticut, the board of health, which as late as 1885 still claimed a role for the "poisonous effects" of "excrement-reeking air," had by 1886 completely gone over to the germ theory.[21] In that year, the board argued that its main function was to "render available and useful the results which bacteriological studies have thus far brought to light."[22] For scientists struggling with the issue of water pollution, the germ theory provided an explanatory paradigm for why disease tended to break out in certain areas. It was not foul-smelling water or poisonous vapors that caused disease. Instead, it was invisible germs passed into the water system and taken in by drinkers farther downstream that caused the spread of disease. By 1891, health activists argued that "experience . . . teaches emphatically, that a sewage-polluted river is not a safe water supply, although some still do not believe this is true for a large river with a relatively small sewage contamination."[23]

Advances in laboratory science, with researchers trained in graduate programs to use the latest technological instruments and chemical and biological analysis, opened the door for new scientific specialists to dominate the field of disease causation and transmission. Increasingly, the generalists on state boards of health stopped merely looking, smelling, and tasting fouled water. Nor did they solicit the testimony of local residents about water pollution. With the germ theory, the smell of the water or its look or taste was not the issue. The concern was whether it had scientifically measurable levels of pollution, particularly the presence of human feces. And human feces were the crucial measure, because they carried germs, particularly the germs that caused typhoid, cholera, and dysentery.

With the germ theory, concern shifted from general pollution to water safety. Water was safe if it did not carry human sewage and hence germs. To determine water safety, or the absence of germs, officials increasingly looked to the reports of professors and specialists such as Ellen Swallow (Richards) and William Sedgwick at MIT. In 1896, Thomas Drown, "Chemist of the Board," reported on the water quality of the Boston Waterworks's reservoirs. In his report, he noted that "a knowledge of the composition . . . is a much surer guide than the unaided eye."[24]

In 1886, when the Massachusetts Board of Health was reorganized, the state also passed the Act to Protect the Purity of Inland Waters, which strengthened the older 1878 act and gave the board additional powers. If water purity was to be protected, and if, as the new germ theory stated, the most significant threat to water purity was germs in human sewage, then the state had to do something about human sewage being dumped into its waters. In 1887, the legislature by special resolve ordered the board to do a scientific investigation on the best method of effecting the sewerage and sewage disposal. This time, the state did not turn to Charles Folson, as it had in 1877, but rather directed Ellen Swallow, a laboratory-trained biological chemist from MIT, to lead the project. Swallow was born in Dunstable, Massachusetts, in 1842. She completed her undergraduate and master's education at Vassar in 1873 and, after being admitted as a special student to MIT, gained her bachelor of science degree the same year. In 1876, she was appointed instructor in the women's laboratory, and in 1884, she was appointed instructor in sanitary chemistry. She was a fellow of the American Association for the Advancement of Science. Swallow developed the world's first water purity tables and established water quality standards for the state.[25] The new standards included the older observational data such as turbidity, sediment, and color but also included dissolved and suspended ammonia, nitrates and nitrites, and oxygen consumed. The data analyzed included the presence of algae, fungi, datomaceae (a borosilicate of calcium), and microscopic animals in the water.[26]

In 1888, with the additional information from the new survey of the state's waters and with the knowledge of the germ theory, the state revised the 1886 law, increasing the power of the state and the board of health over the streams, rivers, and ponds. The new law specifically gave the board power over recommendations for sewers and new systems of pollution disposal. It encouraged the board to report to the attorney general instances of failure of compliance to the state pollution laws. It also gave the board the power to conduct experiments to develop "the best practicable methods of purification of drainage and sewage or disposal of the same. For the purpose aforesaid it may employ such expert assistance as may be necessary."[27]

For Hiram Mills, the hydraulic engineer, Swallow's data indicated that the problem of pollution was an engineering problem, not an enforcement one. Mills and another MIT biology professor, William Sedgwick, set up an experimental sewage water treatment station in Lawrence in 1887.[28] While Mills was a trained engineer, Sedgwick and Swallow were university-trained scholars. After graduation from Yale, Sedgwick briefly attended medical school before switching to graduate work in biology under Henry Martin at Johns Hopkins University and earning his Ph.D. in 1881. In 1883, MIT appointed Sedgwick assistant professor in biology. Sedgwick was interested in laboratory science, and he wanted his students involved in experimental activity. The new experimental laboratory at Lawrence being set up by the state under the auspices of the board of health offered Sedgwick what he wanted.

When the board of health was re-created in 1886, it requested $2,500 from the state to conduct scientific studies of water and pollution. The funds were used to create the Lawrence experimental station under Hiram Mills and to hire specialists to do the work. In 1888, Sedgwick was appointed to the board as its biologist. With Sedgwick as the biologist, Mills as the engineer, and Swallow and Thomas Drown as the chemists, the board began extensive work analyzing the waters of Massachusetts.[29]

The station at Lawrence not only did water analysis but also investigated means of clearing water of harmful material. Health reformers had been interested in ideas of filtration since the late 1860s and early 1870s. It was believed by the early reformers that with the proper attention to filtration, much of what was dumped into the rivers and streams could be either reused in manufacturing or used for manure and fertilizer.[30]

Even before the Lawrence station had proven its success, the New Hampshire State Board of Health raised the alarm about dumping raw sewage into local waters and pointed to the possibility of filtration and purification of municipal sewage. "This mode of disposing of sewage [dumping it into the nearest river] has been accepted as the best, yet . . . common sense as well as common law does not sanction the pollution of our rivers and other bodies of water . . . by delivering large quantities of sewage and mingling it with the current."[31]

The New Hampshire board, expressing American optimism and confidence in science, technology, and business, believed that a solution was sure to be found. "For the time this may be the best we can do, but it will not be complimentary to our inventors and the men who push our business . . . if we do not in a reasonable time find a new way of disposing of waste material and that more in accordance with nature's law that nothing shall be lost." The board went on to contend that "considerable [work] has been done of late to demonstrate the feasibility of sewage filtration with a new [means] of disinfecting the mass and then draining away the fluids and utilizing the solids mixing with loam and using this as a fertilizer."[32] New England public health activists were particularly attentive to the experiments carried on in England on sewage filtration and treatment. But until the 1880s, New Englanders were more impressed with the effort than the results of these English experiments.[33]

Although the reformers continually stressed the possibility of reusing wastes and spreading sewage as manure, little came of their pleading. Cities, towns, and manufacturers found it cheaper and easier just to dump wastes into the local streams. With new means of measuring water quality, and new ideas about filtration, the Massachusetts Board of Health was convinced that its experimental station in Lawrence would finally be able to develop an effective and efficient means to clean water sufficiently enough to no longer constitute a threat to the water purity of the region's inland waters.

Lawrence was the optimal place to locate the experimental station. The city got its water from the Merrimack, where upstream, Lowell dumped its sewage. In 1890, a particularly severe epidemic struck both cities.[34] As deaths from typhoid rose dramatically in Lowell and then later in Lawrence, Hiram Mills launched an intensive investigation of typhoid cases in Massachusetts. Mills discovered that Chicopee, Lowell, and Lawrence had the highest rates. Chemical analysis of water from the Merrimack could not account for the typhoid statistics.[35] For Mills, germs explained the dramatic rise in typhoid, not chemicals and industrial wastes in the water. William Sedgwick of MIT, in his study the next year of the problem of typhoid spread, linked the disease in Lowell with its movement to Lawrence through the Merrimack and those cities' water systems.[36] Mills's and Sedgwick's two studies confirmed the role of germs in spreading typhoid.

Meanwhile, in 1891 and 1892, Hartford experienced a typhoid epidemic. Hartford used the Connecticut River for its water only when its traditional reservoirs were low, which they were in the fall of 1891. The city was concerned because of the pollution Springfield and Holyoke dumped into the river. While the city used the river water, typhoid cases rose dramatically. The investigators for Connecticut's board noticed that just prior to Hartford's use of Connecticut River water, Springfield had a typhoid epidemic in the fall of 1891 that killed thirteen people. The Connecticut State Board of Health concluded that Hartford's typhoid epidemic came from the Connecticut River water. It advised its citizens to boil their water and declared the river polluted and "unsafe for drinking at any point."[37]

Science and Technology's Success

The experimental sewage waste treatment station in Lawrence was a great success. While the English experiments with filtration had limited results, the Lawrence station was able to prove, using the methods devised by Mills, Swallow, and Sedgwick, that bacteria and other germs could be eliminated by filtration through sand and exposure to air and sunlight.[38]

Based on these experiments, the board of health recommended that Lawrence build a filter treatment plant for its water supply. In 1893, Lawrence installed sand filters. According to an investigation by Mills and Sedgwick, the sand filters in Lawrence removed 90 percent of the bacteria in the river water, while mortality for typhoid dropped 40 percent. With this information and the experiments done at the station, health reformers concluded that the old theory that water purifies itself was "insufficient." Typhoid and presumably other diseases were "readily conveyed down stream by sewage polluted drinking water," and it was "practicable and reasonable to protect a community against [an] infected drinking water supply by natural sand filtration."[39]

Following the establishment of the treatment plant in Lawrence, the town of Brockton, Massachusetts, built an even larger and more complex system. Brockton not only filtered its drinking water but also built a sewage purification system. The Brockton system collected sewage in a reservoir, then pumped it to sand disposal fields of some twenty-two acres. The sewage seeped through the sand, leaving behind a thin layer of waste, which was exposed to air and sun.[40] The sewage was then cleaned by natural and chemical action and removed as fertilizer. Lawrence proved the validity of water purification. Brockton demonstrated the possibility of sewage treatment. Whether cities would decide to clean their water before or after they used it remained the question.[41]

Science and technology found a solution to the problem of sewage pollution and the passage of germs into the water supply of the state. The success of the Lawrence station was quickly heralded by public health reformers. Health reform shifted from doctors like Bowditch, Folsom, and Walcott and concerned citizens like Olcott and the residents along Piper's Brook to scientists, chemists, and biologists like Sedgwick, Drown, and Swallow working in laboratories, and engineers like Mills working at experimental stations.[42] The Connecticut State Board of Health argued that not only should boards of health have access to "person[s] versed in the methods of modern bacteriology and familiar with the bacteria which are proven to cause disease," but that "the analysis of drinking water . . . is no longer complete . . . [until it analyzes] the presence of bacteria, upon whose existence or absence may depend entirely the salubrity or insalubrity of the water."[43] "Biological analysis" by trained laboratory scientific specialists was now required, rather than the older viewing, smelling, and tasting.[44]

By the 1890s, the Massachusetts board turned to an investigation of ammonia and dissolved oxygen in water systems. George Fuller and George Whipple, both trained at MIT, discovered that coliform bacteria in water were an indication of human and animal feces because they were not typical water organisms. This discovery allowed for water testing for fecal pollutants and pathogenic organisms.[45] By 1899, the Connecticut State Board of Health was testing water for the presence of "colon bacilli"; although not "pathogenic," the "detection of colon bacilli is regarded as of special significance as indicating probable sewage contamination and the possible presence of pathogenic forms."[46]

Massachusetts's success at Lawrence encouraged other states to establish laboratories to look for bacteria in municipal water supplies. Vermont established a state biological laboratory for "bacteriological examination of water supplies" in 1896.[47] In its investigation of water pollution in 1888, Connecticut focused on both human and industrial wastes. But with the acceptance of the germ theory and the success of Massachusetts's experimental station at Lawrence, health officials in Connecticut now felt that something could be done, at least about human sewage. In 1895, the board reported that "the increasing contamination of certain of our rivers shows conclusively the necessity of providing some other method of sewage disposal than the discharging of crude sewage into the rivers."[48] By the mid-1890s, Connecticut towns were advised to quickly adopt the Brockton or Lawrence station's model system. Pressure to do so was increased by successful lawsuits of downriver residents against upstream polluters.[49] The Connecticut board noted in 1899 that successful suits against cities and towns that dumped pollution were forcing these communities to "find other ways of disposal than the nearest streams."[50] By 1899, the Connecticut State Board of Health believed that "the pollution of our streams is an evil of such magnitude that the attention of the authorities in our inland cities is being more and more directed to the matter of sewage purification . . . [and] for some of our cities there is no other alternative; they must in the near future provide for the purification of their sewage before discharging it into the water courses."[51]

The work of the bacteriologists finally allowed state health officials to have some success in determining if the water that cities and towns were using was safe to drink. The Lawrence experimental station's filtration plants also demonstrated a means by which communities could finally deal with their water supplies. By the end of the nineteenth century, "the use of the larger rivers as direct sources of public water supply without purification ha[d] practically ceased."[52] If the germ theory and modern science and technology could solve the problem of the spreading of disease to downriver drinkers, the germ theory also solved the problem for manufacturers of industrial pollutants. While Bowditch and Walcott complained of industrial pollution in the 1870s and early 1880s, germs now provided manufacturers with a new way to present industrial wastes. Germ theory redefined what was pollution and what was not. Indeed, what Bowditch and Walcott saw as industrial pollution fouling the state's waters, others, using the germ theory, saw as cleansing agents killing germs. If rivers and streams full of fish were signs of health for the earlier reformers, rivers and streams free of fish were a sign of cleaned waters for those opposed to cleaning industrial wastes.

Industrial Wastes as Cleaning Agents

Industrial wastes, manufacturers now argued, helped reduce sewage pollution by killing germs. In Connecticut, scientists for the manufacturers argued that "inorganic chemicals [industrial pollutants] [are] harmless, or positively beneficial in counteracting the organic matter."[53] In 1887, the city of New Britain claimed that the industrial discharges its manufacturers were dumping into the local water were counteracting the sewage in the stream. Although the Connecticut State Board of Health remained skeptical of the claim, it did allow that although sewage pollution was not cleaned by the industrial discharges, it might "be modified by it."[54] An investigation of pollution in the Park River found extensive sewage pollution from sewers dumped into the stream. Despite this finding, an investigator noted that the water below the gasworks was "significantly free from a great amount of bacterial life." The investigator was "not prepared to say whether, or not, the effluvia from these [gasworks were] very harmful," but he did allow that "the odor of the sewage is to a great extent destroyed, and . . . the water is very pure in color."[55]

Earlier environmental health reformers pointed to fish as a sign of water purity. Fish ironically now became suspect, as scientists showed that fish lived in water that also supported germs. "While the impurity of water may in a measure be indicated by its poisonous action on fishes, and while water sufficiently contaminated to prevent the life of fish in it is unquestionably too much polluted for any domestic purposes, it does not necessarily follow that such contaminated streams must be dangerously polluted. Fish will live in concentrated fresh sewage, but will die when the water contains one hundred thousands part of blue vitriol."[56] Indeed, the Connecticut State Board of Health, reasoning from the assumptions of the germ theory, came to the conclusion that industrial wastes do "little harm to the river in a sanitary sense, though they have long since rendered the water of the streams wholly unfit for fish."[57] The germ theorists believed that human sewage, not industrial wastes, caused illnesses. It still remained true that industrial wastes seemed to kill fish, but fish were clearly less important than humans. Industrialists might be held accountable for their wastes if human life was at stake, but fish were a different story.[58] Although many people fought hard to protect the fish in the 1860s and 1870s, as far as health officials were concerned, by the end of the 1880s, people's well-being may have been an important factor in assessing the value of economic development, but not that of fish: "Which are of more importance, fish or the manufactories, there can be but one answer."[59]

With the encouragement of public health officials, gradually, more sewage treatment centers were built.[60] Despite this initial activity, a massive construction campaign of these systems did not materialize. The science that focused on invisible germs also offered cities an alternative health policy to sewage treatment: the option of cleaning the water before it was used and then dumping its used polluted water into the streams, for downstream users to likewise clean their water.[61]

Sewage treatment facilities for wastewater were expensive and benefited not the towns that installed them but downriver users. Cities and towns proved reluctant to install waste treatment plants not only because the expenditure would not directly benefit them but also because of previous sewer line decisions. In the late nineteenth century, when cities began building sewage systems, they had a choice between dual or single systems. Dual systems had separate lines for sewage and rain runoff. On the other hand, single sewers were cheaper to build. If cities were just dumping wastes into the nearest river, the single system was cheap and worked well. Rain runoff flushed out the pipes and diluted sewage. Since many early drains were simply culverted small brooks that ran through the city anyway, and since many of the earliest sewers dumped into these brooks, the expansion of these drains into a single system made sense to pennypinching public officials. Under pressure to build sewage treatment facilities, these municipalities faced a daunting task. They could either dig up and rebuild their entire sewer drainage system to create a dual structure or build a treatment facility large enough to handle the capacity of not only sewage wastes, but other urban water runoff.[62]

Although the germ theory centered attention on drinking water, the question of whether or not to dump used water—sewage pollution—back into the region's streams continued to be debated.[63] With the availability of new sewage treatment technology, the courts moved toward establishing clean water, free of pollution, as a public and individual right. The courts now argued that with the technology of treatment, municipalities "had no right to so contaminate the streams by the discharge of their sewage."[64] More and more cities and towns had to "sooner or later . . . by some means devise [how] their sewage can be disposed of without discharging into the rivers and streams."[65]

Health reformers pushed for cities and towns to adopt "mechanical filtration," which was called the "American system." But cities still balked because of "the high cost."[66] The volume of water in the Connecticut was so great that a study in 1891 found that "the amount of organic matter in the water is not large . . . and only a small part of it can be attributed to sewage . . . The evil results from . . . organic matter, is not what is to be feared from drinking highly diluted sewage. The dangerous elements are the disease germs."[67] If germs were the problem, then cities on the larger water systems argued that filtering the water before use made more sense than cleaning the water before dumping.

Despite the fact that the Massachusetts State Board of Health found that towns with sewage treatment systems were able to treat their wastes so that "sewage is well purified so that it does not have an unfavorable effect on the appearance or odor of the stream into which it flows," larger cities along major rivers resisted building the systems.[68] An inspection in 1902 of sewer outlets into rivers and inland waters of Massachusetts found that only seventeen of the fifty-eight cities and towns it looked at had sewage purification systems. Almost all of these were cities and towns without outlets to large rivers. The larger cities, such as Holyoke, Springfield, West Springfield, and South Hadley on the Connecticut, or Chicopee and Ludlow on the Chicopee, dumped their sewage directly into the rivers.[69]

Dumping raw sewage into the rivers remained objectionable to public health officials, "even though the effect of the sewage upon the appearance of the stream may not be noticeable."[70] In 1902, the Massachusetts State Board of Health reported that "all of the larger rivers of the state are polluted in a greater or lesser degree by the sewage of cities or towns or by manufacturing wastes from factories and mills located along the streams or their tributaries ... [and] in extreme cases the water of the stream is rendered filthy and offensive for many miles."[71] Ten years later, the Connecticut State Board of Health complained that without stronger legislative action and state authority to curb pollution, even the larger "rivers and harbors have reached the limit of sewage pollution so that sooner or later some method of sewage disposal will have to be adopted by the cities and towns responsible."[72]

Failing to get the cities along the major rivers to filter their sewage, health reformers argued that at the very least, cities should make sure that their sewer outlets extended far enough out into the main water flow so that sewage did not accumulate on the shore, and if possible they should have several outlets downstream from the city so that the sewage "mingles ... rapidly with the water."[73] The board of health was particularly concerned about Holyoke because much of its sewage was dumped into the Connecticut just above the dam, which caused waste to gather and flow into the canals and then back into the city. Several of the city's other sewer outlets either were broken or failed to discharge far enough into the river to avoid a noxious condition. The best the board of health could hope for from Holyoke was that the city would rebuild its sewer lines to empty below the dam and far into the Connecticut's main stream.[74] In 1903, the Massachusetts board recommended to the town of Chicopee that it could dispose of its sewage and rainwater runoff by discharging it into the Chicopee River at a "sparsely settled area." It suggested laying the pipes below the water level, fifty feet from the bank. Cities responded to complaints about pollution by building intercepting sewer lines, which cut off sewer outlets to smaller rivers and streams or those dumping sewage close to the city limits.[75] At the same time, the Massachusetts Board of Health realized that although this solution might suffice for the present, "it may become necessary in the future to remove sewage from the stream on account of the use of the water for manufacturing or other purposes below."[76]

In 1909, Hartford established a special committee to investigate ways of reducing "to a minimum ... pollutants in the Connecticut River, and the removal of the increasingly unsightliness and objectionableness from the discharge of crude sewage from local sewers along the immediate river frontage."[77] To accomplish this, the committee did not advise building a sewage filtration system, but rather installing new interceptor lines along the river, which would dump the sewage below the city.[78] The special committee did note that its interceptor line plans would not solve the problem of sewage pollution, but it argued that since "the volume of water in the Connecticut River . . . is so large in comparison with the present discharge of sewage, [further action] can undoubtedly be postponed for many years."[79] The committee noted that when the pollution of the Connecticut eventually required other actions, the city might have to act, but "until then both the cost of construction and operations of [such action] would be saved."[80]

The solution of building interceptor lines and submerged sewer lines well into the main streams of the larger rivers limited the obvious noxious impact of polluted water, but it did not end the problem or justify at the turn of the century the Massachusetts State Board of Health's optimistic claim that the Connecticut River entered the state "practically unpolluted" and left "not seriously polluted."[81] Indeed, in that same otherwise optimistic report, the board also noted that "sewage . . . noticeably pollute[s] the river" where the Millers River joins the Connecticut, as did "sewage and manufacturing wastes from Holyoke, Chicopee, and Springfield."[82] As the Connecticut State Board of Health reminded its citizens, "The construction of sewers does little to solve the question of a sanitary disposal of sewage"; for that, "the purification of sewage by . . . filtration" was necessary.[83]

In an investigation of its rivers in 1908, Massachusetts found that in the Connecticut River "above Northampton . . . the evidences of pollution [were] slight," but that "below Holyoke the effect of pollution by manufacturing wastes is noticeable for a considerable distance . . . and there is a rapid increase in pollution as the river passes Springfield and receives a large additional quantity of sewage." Although the board found that because of the vast volume of water in the river, "the visible effect" of pollution was "slight," yet "the increasing pollution is shown clearly by chemical analysis of the water," and in the Connecticut's smaller tributaries, the pollution was "very objectionable."[84]

For the reformers of the 1870s and 1880s, the science of observation linked industrial and sewage pollution to concerns for public health. Both were clearly observable. The struggle to clean up rivers and streams led to an intractable conflict between economic development and public health. Earle Phelps in his report to the American Public Health Association noted that "experience has demonstrated that . . . if there are no feasible methods for the prevention of a nuisance that nuisance will in general be tolerated, even though it may be a serious one, instead of the industry being destroyed."[85] In saying this, Phelps, who did not support this position, was only reporting what he believed to be a historic truth. That historic truth was of course embodied in the two Merrifield cases, which established that downstream dwellers do not have a right to water in its natural pure state if the upstream consumers were using the water in a reasonable fashion.

Development and Social Costs

New Englanders confronted both the benefits and the environmental costs of the amazing economic transformation that engulfed the region over the length of the nineteenth century. Many people found, in the factories and mills that seemed to spring up at every fall of a New England stream, jobs that were dreary, monotonous, and dangerous, if not deadening to the soul, as Theodore Lyman noted. For many who found employment building mills, dams, and canals, spinning or weaving cloth, pressing paper, or working metal, the work was long and hard, and the wages of the whole family barely put food on the table and a roof over their heads. Yet even many of those whose work life was a living hell felt dependent on the economic transformation that brought the mills and factories. For others who owned or managed the factories, or, like litigation-happy Royal Call, owned tenement houses, or for farmers who could now sell produce to local markets, or for the clerks, merchants, lawyers, and teachers who found employment in the growing manufacturing centers, the economic development provided opportunity. Yet there was a cost to that opportunity. Townspeople were less likely to put out signs saying "gone fishing," for there were fewer and fewer fish in the region's streams and rivers. Life, liberty, and the pursuit of happiness promised in the Declaration of Independence seemed a far-off goal for those living in the crowded tenements, where death rates from dysentery, cholera, and typhoid were several times the rate of rural areas. The great economic transformation that Theodore Lyman celebrated in his run for Congress looked different to the poor. For the poor, even the parks, "breathing place[s] so to speak ... where the weary mother takes her infant ... seek[ing] the shade and the air," had become "open sewers."[86]

Yet despite the costs, economic progress and development were presented as a public good. Many believed the argument that the more mills and dams, the better the region. Others, such as Olcott and Bowditch, argued otherwise. For them, public health and a clean environment were a good just as, if not more, important than economic growth. Those who argued for public health and a clean environment also were able to gain a public hearing and had strong public support. The very jockeying of personnel on the Massachusetts State Board of Health—Bowditch off, Donnelly on; Donnelly off, Walcott on; Walcott off, Walcott on—reflects the failure of either position (economic development or public health) to dominate politics or public discourse. Yet for public health to move forward without opposition, technology and science had to continue to provide means of avoiding conflict between public and private interests.

Public health reformers continued their discourse about clean air, clean water, and clean soil, but they defined these in narrower terms. Clean water meant clean to drink after treatment, clean air meant open spaces between tenement buildings, and clean soil meant garbage pickup.[87] Indeed, at the turn of the century, not only had boards of health become more focused on purification of drinking water than on antipollution, but they also had become more concerned with the costs of clean water and the control of individual use.[88]

Public health activists were aware that industrial discharges were still problematic. In a 1902 report, the Massachusetts State Board of Health admitted that although they were not asked to look into industrial wastes, they still believed that "most of the streams of the state are also polluted by solid and liquid wastes from manufacturing processes discharged from factories and mills." Indeed, "some of the most objectionable nuisances are due solely to such wastes."[89]

9

"Most Beautiful Sewer"

The reforms of the late nineteenth century did help protect New England's drinking water. The plague of water-borne diseases that made the region's cities so infamously dangerous to live in seemed to be in retreat. For a moment, it looked as if the new century would bring a world in which there did not have to be trade-offs between economic development and environmental quality. The ideal articulated by Lyman and Mills—that professional expertise would transcend conflicts of interests between manufacturers and reformers—seemed at hand. Yet there were still problems that these optimists overlooked. And these problems broke into view again in the new century. Despite the health gains, New England's rivers and streams continued to receive massive influxes of pollution of both industrial wastes and human sewage.[1] The larger cities along the major river systems continued their practice of dumping raw sewage downstream, while manufacturers still saw running water as a natural disposal system for their wastes.

Industrial wastes, although less central in the conversation around public health and the environment, were clearly polluting water systems, and reformers never completely gave up the struggle to clean water of industrial pollutants.[2] In its 1896 report, the Massachusetts State Board of Health discussed possible solutions to the problems of "waste liquors or sewage from those manufacturing industries in the State which pollute or threaten to pollute our rivers and ponds." The Lawrence station experimented with different methods of removing industrial wastes. Yet the "problem of successful and economical disposal of this sewage [remained]."[3]

Campaigning for Fish

As people began to look at clean water as an aesthetic as well as a health issue, the ability of water to sustain live fish, which had been dismissed twenty years earlier, now became a concern.[4] Commissions on fisheries that had focused attention on fishways and fish cultivation in the nineteenth century began to revisit the issue of water pollution as they noticed their hatchery fish dying in polluted waters; oyster growers complained to the commissions that their oyster beds were being polluted.[5] Noting in 1905 that "pollution of waters by refuse matter of mills and factories ... is fatal to fish ... [and that] some of the most important streams of this state have been seriously polluted by acids and other poisonous substances to such an extent as to make it impossible to maintain fish life therein," the Connecticut Commission of Fisheries and Game began to lobby for antipollution laws, particularly for laws against industrial pollution.[6] The commission lamented that "there [were] no laws on the statutes which in any way limited water pollution for the preservation of fish."[7] In 1910, the commission went before the legislature to beg for legislation to protect streams and rivers from pollution, noting that "numerous people [had] asked the commission to do something about pollution."[8] In 1911 and in 1913, the Connecticut Commission of Fish and Game two more times asked the state for legislation so that "under no circumstances should anyone be allowed to pollute water in any stream of this state or to make open sewers of them to the extent that will drive out or destroy fish life."[9] In its 1913 report to the state, the commission noted that "pollution of rivers and streams by industrial establishments forms one of the chief obstacles which the Commission encountered in its efforts to effectively establish fish and maintain them."[10] But in each case, opposition from manufacturers and municipalities that were dumping into local rivers and streams prevented legislation from being passed.

The state fishers were not the only ones concerned about industrial pollution. In 1910, the state board of health asked the state legislature for power to stop pollution, but, as with the fish and game commission's requests, the legislature failed to act.[11] Feeling pressure from fish and game clubs, public health activists, and urban reformers, the legislature in 1912 requested that the state board of health investigate the pollution of the state's streams.[12] Again in 1914, the board of health was asked to do a thorough investigation of water pollution. Reflecting the growing interest in clean water as an issue not only of safe drinking water but also of aesthetics, the board of health's 1915 report noted that water pollution could be considered from two viewpoints. One way to understand pollution related to its impact on drinking water, "when in a strictly sanitary sense the presence of pathogenic bacteria in the water and their direct menace to public health is paramount." The board also noted that another way to consider pollution was as a "violation of common decency by the creation of nuisances from the presence in the stream of large quantities of organic matter."[13] Recognizing the accomplishments of past actions toward providing clean drinking water, the board felt that there was no longer a need for special investigation in the first area since most of the state's major rivers were "not used as a source of public water supply without some attempt at purification."[14] The board's investigation focused not on pollution as it might "directly affect the health of any community" but rather on pollution that was "offensive to the sense of decency."[15]

The board of health joined the fish and game commission in arguing that pollution was a public issue because it "threatens the existence of major fish life."[16] In 1914, the board of health reported that "trade waste [was] particularly detrimental to fish life. In many instances they are killed within a short time where large amounts of wastes are discharged; or if the pollution is more gradual, they will be completely driven out in time."[17] The board's interest in water free enough of industrial wastes that fish could survive in it was not motivated by fear that pollution would create a health hazard for those eating the fish.[18] Instead, health reformers were concerned that water too polluted for fish was a nuisance and a health issue in the general sense.[19] On an optimistic note, the 1915 report hoped that "the improvement of these conditions is not impossible nor would the expense be excessive, if remedial action is taken in time."[20]

Water Purifying Itself: The Role of Oxygen and Bacteria

If scientific understanding of the germ theory provided industrial polluters with a means of avoiding dealing with the pollutants that they were pouring into the region's water supply, greater understanding of the chemical processes involved in water purification raised anew the issue of industrial pollution in the twentieth century. In 1910, Earle Phelps, an assistant professor of chemical biology at MIT who had earlier worked at the Lawrence experimental station, argued that because of the role of bacteria and oxygen in breaking down organic matter in water, the appropriate measure of water quality should be the amount of dissolved oxygen (DO) in the water.[21] Phelps noted that wastes took up oxygen from the water and hence put a demand on oxygen, which he called the biochemical oxygen demand (BOD). Phelps's findings quickly spread to other health reformers.[22] By the late teens, chemists and biologists focused attention on the role of oxygen and bacteria as cleansing elements in water purification.[23] Looking at the role of dissolved oxygen and of aerobic bacteria, researchers noted that dissolved oxygen worked with the bacteria to break down organic wastes. Thus, when "the flow of the stream is large, naturally the supply of oxygen is greater, and the oxidizing of the organic matter progresses without producing offensive conditions."

Health reformers then noted that "sewage and manufacturing wastes having been dispersed in the water, the organic matter immediately begins to take up the oxygen contained therein and this proceeds until the organic matter is completely oxidized or until the supply of oxygen in the water is thoroughly exhausted."[24]

Realizing the importance of oxygen and bacteria in breaking down wastes and in naturally purifying water focused attention once again on industrial wastes, as well as sewage. Although industrial wastes did not carry germs, they could overload the natural system of purification by taking up too much oxygen. "When . . . the amount of Oxygen falls below a certain proportion, the organic matter is not completely disposed of, and foul gases and discoloration of the water accompanied by a large mortality among major fish life result." The "foul gases" may not have been a threat to human health, but they were a threat to the aesthetics of the waterways and an offense to the senses.

Reformers also noted that industrial wastes were not only a problem because they overloaded the natural system with organic wastes, but because nonorganic industrial wastes, particularly acids, destroyed natural bacteria that work to break down the organics. "The wastes contain acids [that] . . . have a sterilizing effect and destroy the bacteria [that break down organic matter] entirely."[25] Thus by the middle of the second decade of the twentieth century, industrialists who claimed that their pollutants were not a concern because they might kill germs again found themselves on the defensive. The very process of killing life now made those wastes a threat to the region's waters.

In looking at the level of dissolved oxygen in the Connecticut River, the investigators for the 1915 report noted that as the river left Massachusetts, "the percentage of oxygen, as shown by the analyses, approaches dangerously near the limit necessary to preserve major fish life."[26] In the State of Connecticut, "the river receives . . . pollution from paper and silk mills, . . . distilleries . . . and textile works. At Hartford the river receives practically all the sewage of the city and the wastes from numerous factories."[27] Levels of dissolved oxygen, although low when the river entered Connecticut from the industrial pollution dumped in primarily at Holyoke and Chicopee and from the sewage flowing in from Springfield, dropped even lower after flowing through the factory districts and population concentrations around Hartford. The river had 3.48 parts per million of dissolved oxygen at the state line.[28] Although the amount fluctuated as it passed towns and factories and inflowing tributaries, when the river passed the heavily polluted Park River in Hartford, the level of dissolved oxygen dropped to 2.95.[29] Just south of the Park River at the juncture of the Hockanum River, which was polluted by wastes from mills at Burnside, South Manchester, Manchester, and Rockville, and by sewage from the last three of those towns, the level of dissolved oxygen was at 2.1.[30]

Waterways as Places for Recreation

Concern for recreation (and increase of leisure time) also brought more attention to the region's water. Eighteenth- and nineteenth-century Americans saw rivers and streams as navigational lines or as suppliers of water for drinking, irrigating fields, powering machinery, and providing fish for food. Increasingly in the nineteenth century, they were also seen as sinks to dispose of wastes.[31] But twentieth-century Americans began to see rivers and streams also as sources of recreation, places to fish for pleasure or to go and contemplate nature's beauty. With the growing popularity of bathing in the twentieth century, they were also seen as places to swim. As bathing suits became part of the wardrobes of both male and female middle-class Americans, swimmers also began to look for places to bathe.[32] Where jumping into a body of water was once only a male experience, often done in the nude, now couples and whole families went off to swim and play at the water's edge. Upper-class Americans had long gone to the ocean or the isolated lake-front cabin for recreation. But with the coming of the twentieth century, middle-class and skilled working-class Americans began to look to local bodies of water as places to relax. If they wanted to escape the heat of the city for a swim or a cool relaxing time, it was to the nearby river or lake to which they had to look. In Hartford, a bathing house was built along the Connecticut for changing into swimming outfits. Swimming in the river was particularly popular for lower middle- and upper working-class city residents.[33] By the 1910s, the local Hartford paper ran stories about family outings on the Connecticut River to swim or canoe.[34]

The reduced health crisis of water-borne disease due to water treatment systems reopened the issue of visible water pollution as a nuisance in the early decades of the twentieth century. Fishers, bathers, and canoeists joined public health reformers in agitating against industrial pollution. For those who lived along the inland rivers and streams, the waters to which they looked to fish, swim, or canoe were those very waterways that had become dumps for cities and industry.

Increasingly, these new uses for waterways were at odds with the role of rivers and streams as sinks. River Side Park in Hartford, which had become increasingly popular during the summer as a place to escape the heat for working-class and lower-middle-class Hartford residents, became "unbearable during August because of the smell of sewage in the river."[35] In the summer of 1912, the condition of the river was so bad that the board of health forbade bathing in the Connecticut, despite its growing

Holyoke Canoe Club, Holyoke, Massachusetts.

popularity.[36] Growing interest in recreation increased public impatience with polluted and filthy rivers and streams. That impatience with pollution included industrial pollution.

This concern over water pollution as an aesthetic issue led to a renewed interest in cleaning up the region's waters. Using levels of dissolved oxygen as the measure of water purity, reformers and citizens again began to ask what could be done to improve water quality in the region's larger as well as smaller bodies of water. No longer looking solely at germs, reports about the inability of New England's waters to adequately sustain fish life or maintain a pleasant and scenic setting encouraged reformers and citizens to lobby cities and state legislatures for action to clean up rivers and streams.[37] Agitation by reformers as well as lawsuits from downriver riparian owners increasingly forced cities to begin building sewer systems and, ultimately, sewage treatment centers.[38] Yet most of the sewage treatment centers these cities built did not deal with industrial wastes. New Britain, for example, began building a dual sewer system and a disposal plant between 1902 and 1904, but the system did not connect up most of factories of the area; they continued to dump wastes directly into Piper's Brook, which flowed into the Connecticut River.[39] When industrial wastes did find their way into municipal sewage treatment plants, they caused problems for the plants. Metalworking manufacturers' iron wastes discharged into sewage sewers, for example, clogged the treatment plants and caused filtration systems to malfunction. Reformers argued that manufacturers should be responsible for cleaning their wastes at their site before dumping them either into the municipal system or into local streams or brooks.[40]

Growing Concern for Industrial Wastes

Concern over pollution again put industrial wastes in the spotlight. Citizen pressure led to several failed attempts to pass legislation forcing manufacturers to reduce industrial pollution. At the turn of the century, public support led to the passage of a law in Connecticut to control pollution. Manufacturers lobbied heavily against the law and in 1903 were able to have it repealed. As the fish and game commission noted, the manufacturers objected because they "could not see that the maintenance or increase of trout in a brook is of sufficient consequence to warrant any action on the part of the state which might ... cause them trouble and expense."[41] But public pressure for legislative action did not cease. By 1912, a number of citizen groups and public agencies demanded that the state legislature pass more stringent legislation on water pollution. Farm organizations such as the Connecticut Grange publicly lobbied the legislature to enact antipollution legislation. Joining the Grange were oyster growers concerned that pollution was ruining their beds and fishers concerned about dying fish.[42] The fisheries and game commission asked the legislature for strong laws against water pollution, and the state board of health called for legislation that would give the board power to make "rules and issue orders to prevent pollution."[43]

These boards and organizations had editorial support for pollution control from the state's leading newspaper. The *Hartford Daily Courant* argued in both editorials and articles for strong antipollution legislation.[44] "Summer residents" and common citizens who looked to the local waters for recreation during the hot summer months and were angry that the river was unswimmable because of pollution also demanded legislation.[45] As the *Courant* noted, "The interest of the commonwealth demands a correction of the evil."[46]

Responding to this pressure, the public health and welfare committee of the legislature began hearings on a series of antipollution bills to radically restrict dumping of pollutants into the state's rivers and streams.[47] At the hearings, speakers from a variety of groups and organizations, as well as legislators, testified in favor of the bills. Despite what appeared to be widespread enthusiasm for the bills, they were reported out of committee unfavorably.⁴⁸ Yet public pressure would not let the issue die, and despite attempts of opponents to kill the bills in committee, an antipollution bill was brought back to the floor of the legislature for debate. Again, several representatives spoke out in favor of the bill, particularly those from Hartford and Stamford. But support was not universal. Representative Healy from Windsor Locks, who regularly opposed legislation he felt was detrimental to the manufacturing interests of the state, argued against the bill, along with Representatives Alcorn of Ensfield and Hull of Clinton, both of whom had a history of supporting the Manufacturers Association.⁴⁹ Following intense discussion, Representative Hull, in a parliamentary maneuver that caught the bill's supporters by surprise, had the bill amended in a way that stripped the board of health of its enforcement power, leaving the board with the power only to investigate pollution. This amended version of the bill was sent to the senate, where it died with the end of the legislative session.⁵⁰

Although the bill had widespread public support, it did not have the support of the Manufacturers Association. The manufacturers not only had the support of a number of influential legislators, they also had an extensive lobbying system in place. The combination of intense behind-the-scenes lobbying and supportive legislators enabled the manufacturers to amend or kill antipollution legislation without ever having to publicly declare their opposition.⁵¹

The amended and weakened antipollution bill, modeled on legislation already passed in Massachusetts, was reintroduced and passed in 1915. It restricted the disposal of sewage in inland and tidal waters and charged the state board of health with the task of providing information and assistance in installing municipal sewage systems and disposal plants.⁵² This bill did not address industrial pollution. Nor did it satisfy the pollution reform forces, who came back to the state legislature again asking for an even more sweeping antipollution bill in 1917. The proposed 1917 bill was directed at industrial pollution. It suggested prohibition against the dumping of industrial wastes and penalties for violations.

In the 1917 battle to defeat what the Manufacturers Association considered a "drastic bill on waste disposal," Connecticut manufacturers realized that they had to find a more cooperative approach or else face even more significant legislation. In fact, the Connecticut Manufacturers Association suggested "cooperation between the state and Connecticut manufacturers."⁵³ As a result of conferences between the manufacturers and Connecticut State Board of Health officials, a new bill, partially written by representatives of the Manufacturers Association, was offered as an alternative to the initial 1917 bill. With the support of the reformers and without opposition by the manufacturers, this bill passed the legislature and was signed into law by Governor Marcus H. Holcomb. The new act created a special commission of the board of health that tellingly named itself the Industrial Wastes Board and focused on "such problems as referred to conditions resulting from industrial waste."[54] This board was composed of members from the state board of health, with the addition of two manufacturers, two engineers, and one representative of the state at large.

Governor Holcomb appointed as one of the representatives of the manufacturers Ernest Wilson Christ of New Britain. Christ was the secretary of the Stanley Manufacturing Company and manager of its pressed-metal division. As an officer in the Stanley Manufacturing Company, Christ had more than just a general interest in how the industrial wastes board would proceed. His company had been singled out by the board of health since the 1880s as a notorious contributor to industrial wastes that eventually made their way into the Connecticut River.[55] After the passage of the law, "a number of manufacturers signified their willingness to cooperate with the new Commission . . . for the successful development of methods for recovering valuable by-products."[56]

The board approached the problem of industrial wastes along two lines. One was "the study of wastes by classes . . . for intensive study and experimentation with the idea that the solution of the problems surrounding any particular class of wastes could be applied . . . to all industries of that class." The second was "the study of certain areas or watersheds which are now grossly polluted, for the determining satisfactory methods of relief."[57]

The industrial wastes board hoped that, through study and experimentation, plans could be "found that many wastes which have been polluting the streams can be reclaimed for a profit."[58] It was an old hope that would not die. It was also a hope that manufacturers encouraged. Manufacturers expressed a willingness in "cooperat[ing] with the Board" in order to make the "work [of the Board] a success."[59] Manufacturers had an interest in this cooperation, especially in working with the commission on developing successful "methods for recovering valuable by-products." As long as the focus of attention remained on finding ways to profitably reduce pollution, manufacturers hoped they could forestall legislation forcing them to assume immediately the costs of pollution reduction.

The board also hoped that by working with manufacturers to find ways to make pollution reduction profitable, a prolonged political and economic battle with the manufacturing interests could be averted. It was a battle that, if engaged, could have significant impact on the state. In its investigation of the Hockanum River, the industrial wastes board found that paper and textile manufacturers were the primary polluters. The board also noted that these companies employed either directly or indirectly 75 percent of the watershed's population. Just ten years earlier, the Connecticut Commission of Fish and Game, in looking back over the failure of the state to control industrial wastes, noted that solving the problem was "exceedingly difficult" because it pitted the "owners of industrial plants" who resisted pollution cleanup against "the rights of the people to have their health safeguarded . . . and the rights of people to . . . fish." The commission noted that manufacturers argued that "the output from an industrial establishment" and "considerations of capital invested in industrial enterprises" were far more vital than "the maintenance of fish life." Using these arguments, the manufacturers were able to have "measures . . . looking toward the abatement or abatements of water pollution . . . either . . . defeated or amended as to make them practically inoperative."[60]

Cooperation with the manufacturers to reduce pollution was far preferable to the board members than the possibility of the loss of industry and jobs, or the defeat of pollution abatement measures in the face of manufacturers' intransigence.[61] As Earle Phelps of the American Public Health Association noted at the turn of the century, antipollution legislation that forced manufacturing interests into costly actions to clean up pollution could be counterproductive. Such actions, Phelps argued, would leave industry in the state that passed the legislation at an unfair competitive disadvantage.[62]

The Connecticut Industrial Wastes Board looked at several cases of industrial pollution. The board investigated specific industries and companies and optimistically reported on its successes at cooperating with industry to reduce pollution. Pollution of the Noroton River from oil from the Stamford Rolling Mills was significantly reduced when the board was able to suggest ways of recovering the oil "with a profit." The Connecticut Chemical Company was also persuaded to stop dumping profitably recoverable chemicals into the local stream.[63]

The board, although committed to working with industry, still argued that "the economic value of a stream is greatly reduced or destroyed by gross pollution . . . and that domestic and industrial wastes should have their objectionable qualities removed before being discharged into the stream."[64] The board also felt that "every effort should be made to prevent additional pollution of streams." Maintaining the hope that cleaning up industrial pollution could be profitable, the board reported to the state that "in removing objectionable materials from the wastes of industries of nearly every class, sufficient valuable materials may be recovered to make the cost of removal reasonable and in some cases profitable."[65]

The industrial wastes board did pursue industrial polluters. It also attempted to cooperate with industry to avoid conflict. When the board was concerned about wastes from paper manufacturers polluting Lydall Brook, the board met with "the manufacturers at which cooperation with [the board] was promised in investigating as to best methods of disposing of these particular wastes."[66]

Cooperation was more often promised than delivered. The Emery Wheel Factory in South Meriden, for example, had "several times promised they would install proper machinery to eliminate the nuisance [of smoke and dust]," yet little was done.⁶⁷ Or in other cases, such as the Aluminum Smelting Company of Hamden, in which companies took antipollution action, the results "proved unsatisfactory." In such cases, the board was forced to suggest that local communities take legal action, or the board simply shrugged its collective shoulders in dismay.⁶⁸

The industrial wastes board worked to get companies to develop alternatives to dumping and, in the spirit of the original agreement between the board of health and the manufacturing interests, concentrated most of its activity on working with industry to develop profitable means of reducing pollutants. Looking at the wastes in the metal industry, particularly acids and metal filings, the board investigated with the Case Metal Works in Waterville the electrolytic method for recovering copper and zinc from sulfuric acid. The board found that "careful cost estimates indicate that copper, zinc, and sulfuric acid could be recovered profitably from wastes" at some plants. Also investigated, with the cooperation of the Stovill Manufacturing Company and Christ's Stanley Works in New Britain, were means of recovering acids from dips used by manufacturing companies.

Yet despite progress in experiments in finding profitable means of reducing pollution, at the end of its first year, the board had to conclude that "the proportion of metals and acids kept out of the streams [profitably] is not sufficient to warrant their adoption on the basis of the stream improvement." The board argued that since "profitable treatment and effective stream protection" were not yet available, "it would be unwise to advocate the adoption of any definite method of treatment at the present time."⁶⁹ With such a conclusion, it is not surprising that a "spirit of cooperation [was] manifested by the manufacturers." Optimistically, the board suggested that the manufacturers realized "more clearly the necessity for controlling the discharge of water polluted by wastes from a plant."⁷⁰

Holding true to the belief that science and technology would eventually find a way to profitably remove wastes, the board advised more study. A year later, in 1920, the board again reported that although electrolytic or chemical precipitation could work to profitably remove some pollutants, "definite recommendations should be deferred until other processes have been studied."⁷¹

Given the assumption by the industrial wastes board that the cleaning up of pollution should be profitable or postponed, the board again found the manufacturers cooperative. The more problematic conclusion of the board was that "the extensive equipment and technical assistance supplied . . . represents very definite evidence of their [the manufacturers] interest in the problem of stream pollution."[72] Having accomplished so little, the following year the board was abolished.

Although the board ceased to exist, concern over pollution in Connecticut did not. On May 27, 1923, the state assembly created a new antipollution agency, the Connecticut State Water Commission, to carry on the work of the industrial wastes board. Reflecting the continued public interest in abating industrial pollution, the commission was empowered to eliminate pollution and to prescribe the means by which it could be reduced.[73] Yet despite the legislative mandate, intense lobbying by manufacturing interests and the continued belief by the commission members that a profitable means could be found to reduce pollution led the commission to focus on investigation rather than enforcement. Believing that clean water was needed for manufacturing as well as for recreation, the commission hoped that industrialists would cooperate and avoided taking action against them.[74]

Public Interest versus Private Interest

In 1905, the Connecticut State Board of Health stated quite boldly that "as condensation of populations in cities goes increasingly on, the supply of the necessities of life become more and more matters of public concern and less and less within the direction and knowledge of individuals."[75] The board of health also noted the potential for conflict between private and public interests, especially with rapid economic development; "a business enterprise . . . [is] with some exceptions more concerned for the profits of the stockholders than for the purity of the wate. . . . Private greed is a serious menace to the public health." The board was concerned over "the antagonism, at many points, of the interests of money-makers and the interests of public health." Given what the board perceived as the "opposing interests" of private corporations and the public, it believed that "there will be constant need of vigilance and a judicial sanitary administration to counteract this ever-living and active tendency to increase profits of commerce."[76]

If the Connecticut Board of Health wished to emphasize that private economic interests and public health were in opposition, others, including later members of that very board, focused on how technology and science could overcome that opposition. Indeed, progress was made in the area of public health. Even the concerned Connecticut State Board of Health of 1905 noted that "public attention is now directed as it never has been before to the means of preserving, protecting, and elevating the standard of public health."[77] Public vigilance, state action, and the advantages of modern technology had led to a condition where by 1904, "the cities of Connecticut [were] as salubrious as the country."[78] To achieve this progress, state boards of health needed "scientific assistants and . . . chemical and bacteriological laboratories," but they also needed the authority of the state to guarantee that the citizens all had access to "pure air, pure water, and pure soil, . . . the cardinal or pivotal conditions more favorable to public health."[79]

When public health officials began to revisit the issue of industrial wastes and water quality generally in the region's rivers and streams, they were mindful of the battles of the 1880s. And if they forgot, the serious opposition in Connecticut to a strong antipollution law was a clear reminder. Hoping that technology and science could provide a means of transcending a potentially intractable conflict between economic development and a clean environment, Connecticut's health reformers jumped at the opportunity to work with the manufacturers. The manufacturers conditioned their cooperation on their representation on the board and on the premise that antipollution actions would be predicated on either profitable alternative actions or ones that would represent at most a "moderate expense."[80]

Connecticut's health reformers accepted these conditions in order to avoid continued conflict with the manufacturers. They also accepted them because fundamentally they shared the belief and hope that technology and science could produce a cost-free clean environment. Yet despite their optimism, ultimately the Connecticut reformers found, with some exceptions, that they could not successfully clean the rivers and streams of their state in such a fashion as to be cost free. In the end, the reformers were frustrated in their attempt to find such a solution, even with the cooperation of the manufacturers.[81]

To effectively deal with industrial wastes would cost, no matter which technology came forward. There was no free lunch, nor, as Joel Tarr notes, an ultimate sink.[82] Yet conservationists continued to hope for such a conflict-free environmentalism. If science and technology could be harnessed through the agency of the state to provide both a clean environment and environmentally cost-free economic development, then the state could act as a neutral agent for the benefit of all. If, on the other hand, such a conflict-free situation could not be orchestrated, then the state with its ancient charge as the agent of the common good would become an instrument by which, through democratic struggle, decisions would be made concerning what is a good and for whom. This realization, although seldom articulated, underlay much of the debate over state agency. It went to the heart of the Progressives' vision of the new democratic-activist state. It informed not only the conservationists like Gifford Pinchot and Theodore Roosevelt but other Progressives as well, such as Louis Brandeis, Jane Addams, and John R.

Commons. It was a vision that believed it could hold in harmony the ideal of science and technology transcending conflicts of interest and the role of the state as arbitrator for the larger public good. Unfortunately, these two beliefs were not easily conjoined. The search for a scientific and technological transcendence could mask conflict and undermine the ability of the state to perform its other role as agent of the common good. This was the world that Henry Bowditch understood. His struggle for state medicine did not ultimately create the agency for the common good that he envisioned, but it certainly created a context within which we are still immersed.

10

Farmers, Fishers, and Sportsmen

At the end of the nineteenth century, Edward Bellamy, one of the Connecticut River Valley's most famous literary residents, created a fictional character who wanted to avoid "industrial existence" and instead "all day to climb these mighty hills, feeling their strength" and to "happen upon little brooks in hidden valleys." Bellamy planned for his protagonist "to breathe all day long the forest air loaded with the perfume of the forest trees."[1] The wanderings of this turn-of-the-century fictitious character through thick forests and deserted hills reflects the changes engendered in the valley with the coming of industrial cities and the abandonment of hillside farms. When Bellamy was born in 1850 at Chicopee Falls in western Massachusetts, the region was in the process of deforestation and had few areas that were not intensely farmed. Yet as Bellamy himself noted in an 1890 letter to the *North American Review*, "the abandonment of the farm for the town" had become all too common.[2] Deserted farms became one of the themes Bellamy sketched out in his notes for the novel. Bellamy had his character live in an "abandoned farmhouse. . . . The farmhouse was one of the thousands of deserted farms that haunted the roadsides of the sterile back districts of New England."[3]

In viewing the depopulated countryside as a retreat from industrial existence, Bellamy's character represented the fate of late-nineteenth- and early-twentieth-century New Englanders. Increasingly, urbanized New Englanders began to look to rural areas not as sources of food or resources of necessity but as places to contemplate nature and practice fishing and hunting as sport. As rural areas, particularly on the hills and up the valleys, became less populated, farmers there lost much of their political voice. New city voices now became more important in the conversation about resource conservation. What farmers saw as abandoned and ruined farms, urban and suburban naturalists saw as rural retreats from the tensions and pollution of the cities.[4] For these interlopers, rural New England represented a romantic ideal of a past they or their ancestors put behind them when they moved to the city. And these new nonrural naturalists acted to protect these retreats in the terms in which they saw them.

Old Farms and New Farming

As New England industrialized, farms on the rich alluvial soil beyond the cities grew prosperous shipping vegetables, fruit, and dairy products to hungry urban populations. But the place of the marginal hill farmer became more tenuous. Increasingly, the hilly countryside was deserted by farmers whose families had settled those areas when Timothy Dwight was making his trips.

The story of failed farms on the upland hills and in the isolated valleys and the slow return of pastures and gardens to forests characterizes much of rural New England from the middle of the nineteenth century onward. A recurrent theme in the regional farm journals was the abandonment of farms by farm children, as they headed to the cities and the West.[5] More than half the towns of Vermont (mostly in the upland regions) were losing population by 1860.[6] What was happening in Vermont was also happening in the upland areas of New Hampshire and Massachusetts. Hampshire County, Massachusetts lost 33 percent of its population in the second half of the nineteenth century, while the state as a whole doubled its number of inhabitants. Two-thirds of New Hampshire's townships and three-fourths of Vermont's declined in population between 1880 and 1890. The *New England Farmer* lamented in 1871 that "while the prospect is thus one of great natural beauty, it is, at the same time, one of desolation and ruin, agriculturally.... These families have disappeared, and their farms have become pastures and wood lots, or bare ledges."[7] The "perfume of the forest trees," as Bellamy put it, did indeed replace the smell of manure and fresh-turned earth.

Dairy Farming

Not only was rural New England losing population, but the nature of its agriculture was changing. To survive, many northern valley farmers switched to dairy farming in the second half of the century. Dairy farming, for those near the urban centers such as Holyoke, Chicopee, Springfield, or Hartford who could easily deliver fresh milk daily to waiting urban mothers, provided a good source of revenue.[8] Fortunately for northern dairy farmers, the growing demand for fresh milk in expanding urban areas soon outstripped the supply from surrounding farms, and milk buyers came farther north for supplies. By the second decade of the twentieth century, northern New England's milk production actually increased, with Vermont leading the nation in per capita dairy production.[9]

While northern New England farmers were trying to hang on by switching to dairy farming, in the Connecticut River Valley to the south in Massachusetts and Connecticut, farmers focused on specialized market crops. Although farmers close to the urban areas continued bringing general perishable garden-crop produce to the bulging populations of workers, clerks, homemakers, and children, other valley farmers found in tobacco and onions crops with a high enough return to maintain their farms.[10]

By specializing in dairy farming in the north or by concentrating on tobacco and onions farther south, some New Englanders continued farming, but they were not continuing a way of life that stretched back to their great-grandparents. Commercial farming, which began in New England in the early nineteenth century, became ever more specialized as the century wore on. By the early years of the twentieth century, the reach of the city had transformed the countryside into a rural image of itself.[11]

In order to be successful, these farmers also worked the land significantly differently than their great-grandparents had. Fields were no longer left fallow for a year or more as pastures. Cultivated fields were kept in production yearly, with ever greater amounts of manure dumped on them. By 1920, central valley farmers were collectively buying train-car loads of manure from dairy farmers farther north. In the early spring mornings, wagons could be seen heading out into the fields loaded with manure to be shoveled out onto freshly plowed land.[12] Late spring freshets washed fertilizer from the fields into the Connecticut and its tributaries, adding nitrates (biochemical oxygen demand) to already pollution-burdened waterways. By 1920, acre upon acre of the Connecticut River Valley floor from Northfield south to Hartford, except for the urban areas around the mills and factories, was turned and planted.

Harvesting the Forest

But farmers farther up the hills had neither the capital resources nor the fertile soil to make tobacco or even dairy farming a profitable enterprise. Those farmers who continued to eke out a living in the hill country looked to their woods for survival. With the return of forest cover to the abandoned hill farms toward the end of the nineteenth century, many of the remaining farmers harvested wood and wood products to survive. For generations, each spring before the snow had even melted from the hillsides, northern farmers had gone out to the woods to tap sugar maples for sap, which they would boil down into syrup for their breakfast tables or make into maple sugar candy. Increasingly, those northern farmers found in the sap of the sugar maple a source of income.

By the turn of the century, the growing popularity of New England rural resorts became a major source of revenue for northern New England communities.[13] Summer lodging houses not only sold local crafts and maple syrup but also bought agricultural produce from local farmers, which kept many farms from reverting to forests. Resort hotels, commodifiying the virtues of rural New England, advertised themselves as places where city families could find peace and quiet contemplating a picturesque landscape and quaint (but partially deserted) country villages. City dwellers hoping to recapture a lost but romanticized past flocked to these expanding resort complexes, which employed large numbers of locals, especially children, as servants, cooks, livery operators, guides, and handymen.[14] With the coming of the automobile, twentieth-century tourists who came to see fall colors found village markets and roadside stands full of apples, cider, and maple syrup.[15]

Rich in Habitats, Poor in Game

If some of New England was returning to forest, the region's fauna was slower in reverting to its original fecundity. Bellamy had his fictitious character figure that he could shoot game and catch fish from "the wood and trout brooks around."[16] This remained possible only in fiction. Despite the return of some forest, fish and game remained scarce. Fish, particularly, were in short supply. As one writer to the Connecticut Commission of Fisheries and Game stated in 1909, "Everywhere, the country over, streams once yielding bounteous supply of fish have long since been utterly and permanently ruined by having been converted into sewers for refuse which destroys life."[17]

The attempt of the state fish commissions to deal with the problem of declining fish populations had limited success.[18] Shad, which the fish commissioners feared in the 1860s were on their way to extinction in major New England rivers, increased dramatically in the 1870s and early 1880s.[19] These successes were testaments to the optimism of the early fish commissioners and their belief in the possibilities of science and technology, particularly the fishways and the fish cultivation projects.[20] Following Massachusetts and New Hampshire, Connecticut set up its own successful fish hatchery.[21] By the end of the century, millions of shad fry were planted in the Connecticut River, and the numbers of shad caught by fishers jumped accordingly.[22] However, because of conflict between Connecticut and Massachusetts and the continued obstruction of dams, most of the increase in shad was confined to the waters below Enfield.[23]

As the fish commissions invested more energy in fish hatcheries, they remained frustrated by the ever growing pollution of the rivers, which undermined their efforts. "Many of our streams that were once the habitat of the brook trout, whose judgements as to the purity of water is as good as the results of a chemist's analysis, are simply open sewers exhaling disease and death on their journey to the sea."[24] In 1914, the Connecticut Commission of Fisheries and Game again reminded the state to notice "the matter of damage to fish caused by pollution of the streams of the state, and urge such legislation as will eliminate the present condition."[25] In 1911, the commission complained again about the frustration of having its restocking programs undermined by industrial pollution. "Pollution of rivers and streams in Connecticut by industrial establishments forms one of the chief obstacles which this commission encounters in its efforts to effectively establish fish and maintain them. No authority is conferred by law upon your commissions to interfere to prevent any one from using the best trout or bass streams after having been intensely stocked with fish, to drain deleterious substances into [them]."[26] And in 1915, the fish commission complained that "at the present time some of the waters of our state that once contained thousands of fish are practically depleted because of pollution."[27] Focusing particularly on industrial wastes, the 1915 commission report noted that "some of the most important streams of this state have been seriously polluted by acids and other poisonous substances to such an extent as to make it impossible to maintain fish life therein."[28] The trout brooks, from which Bellamy's fictitious character had hoped to catch his dinner, were by the early twentieth century "used as open sewers by industrial establishments."[29]

Concern over water pollution by the supporters of fish restoration brought them into conflict with manufacturers who were already battling the board of health over antipollution legislation. As the Massachusetts Commission of Fisheries noted as early as 1874, "The prejudice of mill-owners . . . has more or less retarded and increased the labors of the commission."[30] The supporters of restoring fish to the region's waters realized that manufacturers represented a powerful economic interest that could not be ignored, but they believed that water clean enough for fish was also an important public interest.

New technology in the twentieth century complicated the tasks of the fish commissions in returning anadromous fish to the great rivers of New England. Increasingly in the twentieth century, most mills shifted to steam. Waterpower sites went unused, and dams, as Bellamy noted, "returned to a state of nature."[31] With the development of electrical power in the late nineteenth century, however, industrialists and the new electrical power entrepreneurs took renewed interest in these sites. The early-twentieth-century electric power companies began converting waterpower sites into electric generating facilities.[32] In 1909, the Holyoke Water Power Company was authorized to convert its large dam at Holyoke to provide electric power generation.[33] Increased electric power demand led to the creation of the Connecticut River Power Company, which built a hydroelectric dam across the Connecticut in southern Vermont, with turbines with 28,000 horsepower. And on the Deerfield River, a series of dams and reservoirs was built to generate electrical power.

Initially, these hydroelectric sites provided electricity for local cities, usually the industrial cities that had grown up around the original dams. Local communities, anxious for cheap electric power, supported dam rebuilding and new dam heights.[34] Increasingly in the 1920s, the dams generating electric power came under the control of the utility syndicates. Outside interests' control of the region's electrical power sources came to be a serious political issue in New England. Hostility to this situation led to the attempt of the Massachusetts Public Service Commission and the reform Republicans in Vermont to control the growing power of the utility holding companies.[35]

The battle over utility companies' monopolization of electrical power sites in New England, and the cost of electricity, soon superseded concern over the impact of the dams on the local environment. It proved almost impossible to get the new electrical utility companies to respond to the needs of farmers and local businesses for fair rates and electrical service. The possibility of having the utility syndicates respond to the interests of fishers was even farther out of reach. New England fish and game commissions began looking in other directions for help.

Game Protection: "Public Spirited Men of Large Affairs"

Increasingly, the commissions came to see their success not in overcoming the resistance of the manufacturers but in avoiding it. That avoidance was also furthered by the class background and assumptions of the fish commissioners. By the end of the century, the commissioners were almost all upper-class gentlemen who served the commission on a voluntary basis. Most were also naturalists and sportsmen.[36] Frederic Walcott, president of the Connecticut Commission of Fish and Game from 1925 to 1927, was a highly prosperous textile manufacturer and an investor in power companies and utilities. After college, Walcott moved into his family's textile business. In 1905, he became president of the New York Mills and the Walcott and Campbell Spinning Company. Walcott lived in Stamford,

Connecticut, in the New York City suburban belt. He branched out from textiles into investment banking and electric power companies, particularly hydroelectric. Walcott was also an amateur ornithologist and an avid sports hunter and fisher. He was elected to the Connecticut State Senate in 1925, and in 1928 to the U.S. Senate. Like Walcott, the other members of the fish and game commission were, in their own words, "men of the highest character, many of them well known throughout the state as public spirited men of large affairs."[37]

In the early battles over restoration of fish to the region's waters, different groups of fishers and supporters of the restoration of fish cooperated. Commercial fishers made up one such group. The advocates of the poor and the rural Arcadians who believed that the return of nature's abundance would provide the poor with cheap, quality nutrition made up another group. They believed that hunting and particularly fishing were important parts of country life, not only providing food but also linking the farmer to a more idyllic past. Sports fishers made up a third group. Although these different interests supported the return of fish to the rivers and streams of the region, they also at times conflicted with the fish commissions over net size, fishing restrictions, and catch limits.[38]

The conflict over restrictions on fishing went back to the earliest colonial times in cases that came before the early courts. It also was intertwined with assumptions about the cause of the decline of the fish in the first place. The decline of New England's fish, as we have seen, was rooted in four factors: overfishing, obstructions to migrating fish, pollution of water, and destruction of spawning grounds. Overfishing took its toll on the fish because their populations need a certain level to sustain themselves. If overfished, the population will crash.[39] Dams prevented fish from getting to their spawning grounds or kept young fish from returning to the sea. The corruption of the water by industrial and sewage pollution killed the fish or their food source directly or depleted oxygen in the water so that the fish could not survive. Spring spawning and the spawning grounds destruction caused by excessive flooding, and the silting up and corruption of the spawning beds and fish streams due to the deforestation of the region, also contributed to the decline of New England's fish.

In the debate over the restoration of fish, these interacting factors were usually considered in isolation. In the initial meetings of the New England Commissioners of River Fisheries, "great differences of opinion prevailed as to the chief causes of the [decline of fish]. Some asserted that in the Connecticut [River] it was due solely to the pounds [pound nets]. Others declared that the [decline] was due mainly to the small mesh of the nets. . . . Others asserted that the dams alone were the cause."[40] Each of the problems required a different solution. By focusing on the problems in isolation, however, the supporters of fish restoration often were at odds with each other and failed to develop a comprehensive approach.

The problem of dams and of industrial pollution forced the supporters of fish restoration to confront the manufacturers. The Connecticut commission noted in its philosophical overview of the problem of fish depletion, "With an eye solely to manufacturing interests, man has also excluded some of the best species of fish from the spawning grounds."[41]

The problem of overfishing, although real, focused attention on the fishers and often was framed in class terms. In 1896, the Connecticut Commission of Fish and Game noted that "the [nonsport] fishermen, as a rule, have given intelligent thought only to the methods of taking fish . . . [and] whether or not these methods will exterminate the fish themselves in one or twenty years is a question about which they do not generally concern themselves." The commission continued, "Without restrictive legislation there is hardly room to doubt that all the edible or salable fish of this state would be exterminated. . . . There are some fishermen who realize the situation . . . , but there are still far too many who claim that the wholesale present destruction is the legitimate and only safe method of insuring a large future supply."[42] In the twentieth century, the commission continued to complain to fishermen about overfishing. "The interests of the public at large can not but be held to outweigh the selfish interests of a comparatively small number of fishermen whose principle is to profit by the slaughter of the spawning shad."[43]

Although shad numbers were drastically cut by dams and by pollution, those problems proved far more intractable than the problem of overfishing. Overfishing could be blamed on the fishers themselves.[44] The focus on overfishing also went to the root of assumptions about civilization and resources. As the Connecticut commission noted in 1906: "The rapid decrease in the supply of fish . . . as observed from year to year, was a cause of serious alarm to man, but was generally regarded as, in some mysterious way, the necessary and inevitable result of civilization. The true cause was man's ignorance and thoughtlessness. Man's pursuit of fish, whether for food or sport, had always been reckless. Ignorant of the nature and habits of . . . fish, . . . he has destroyed them." Despite their view of man's history of reckless overfishing, fish and game commissions remained hopeful. "But for the wonderful fecundity of fresh water fish our rivers and small lakes and ponds would have been entirely depopulated. Laws [to protect fish supplies] have been recognized."[45] Yet even with these new laws, the commission also noted that "the spirit of fishermen . . . seems to have rebelled against fish laws and they have generally remained a dead letter on the statute books."[46]

Sportsmen Conservationists

The debate about fish restoration, for the last half of the nineteenth century and the early years of the twentieth, focused on the need "to restore food fish to their old haunts," as "food for our people."[47] It also encompassed an economic utilitarian rationale that the nineteenth-century commissioners worried was not appreciated: "Neither the fisherman nor the public realize the value and economic importance to the state of the fishing industry."[48] The fish that occupied center stage in this discussion were shad and salmon.

Yet the argument for fish restoration also encompassed other interests than just fish for food, as was reflected in an 1897 report from the Connecticut Commission of Fisheries and Game. Restoring fish, the commission argued, would provide "cheap healthful and abundant food for all classes of its people, the livelihood for the hardy fishermen, the mental relaxation for the toiling masses of our cities and towns in all the busy pursuits of our intense and intensifying daily life." Fishing as a means of supplementing the diets of the poor was joined here with the argument for its aesthetic value as "mental relaxation" for urban residents. The commission defined those urban residents as "toiling masses," yet in practice, those who fished for "mental relaxation" were more likely to be the urban elite. Nonetheless, this nineteenth-century commission did not abandon its traditional concern for fishing for food but rather merged that ideal to the newer one of fishing for enjoyment and sport: "These and many other objects of vast importance, socially, morally, and financially, are involved in trying to preserve and maintain for the enjoyment of the present and future such food fish as are adapted to our waters."[49]

Although these nineteenth-century commissioners argued that fishing for sustenance and fishing for pleasure were linked, other supporters of fish restoration were less concerned about fish as food than about fish for sport. These were the sportsmen.[50] The sports fishers supported the activities of the fish commission because they were interested in the commission's work of stocking ponds, lakes, and trout streams. Sports fishers were not as interested in shad in the great rivers, such as the Connecticut, as they were interested in trout, land-locked lake salmon, pike, or bass. For most of the nineteenth century, these sportsmen kept a low profile and used the language of fish for food and the needs of the poor to support the activities of fish restoration.[51] Although the sports fishers articulated their concern in the language of restoring fish as food to the poor and restoring abundance to rural Arcadia, they in fact had different interests than the poor or the marginal farmer. These sportsmen were for the most part urbanites. They lived, like Theodore Lyman or a commissioner of fish and game from Connecticut, Frederic C. Walcott, in the cities or suburbs. Shad and salmon on the major rivers provided fish in abundance for the poor, the marginal farmers, and the commercial fishers. The sports fishers were interested in the chase and the wilderness, or at least "natural experience," and the thrill of the catch. The fish they were interested in were hard to find and difficult to land. The aesthetics of the catch counted for them.[52]

Industrial water pollution was of greatest threat to the fish, particularly shad, in the major rivers; it posed no threat to the fish in the distant trout or bass streams and lakes. Increasingly, as the New England commissions on fish and game shifted their attention toward sports fishing, they found themselves in less conflict with the industrial interests. The shift to sports fishing and hunting also involved a move away from the traditional justification for fish restoration as "cheap food for the people" so that a "poor man with his quarter may obtain a good meal."[53]

By the twentieth century, what had been the old commissions on inland fisheries had changed their names to fish and game commissions, and although their focus shifted toward sports fishers and hunters, the newly named commissions retained elements of their earlier discourse. In 1914, the Connecticut Commission of Fish and Game noted that one of its purposes was to ensure the rights of the people of the state to hunt and fish near their homes, for the "wild fish and game of the state belonged to the people and not the fortunate few."[54]

The increasing focus of the fish and game commissions on the interests of sportsmen raised opposition among people who saw this activity as benefiting the elite rather than the common citizen. The charge put the commissions on the defensive. "Sometimes it has been charged by those who have not given the subject careful consideration, that this commission is largely engaged in propagating game fishes for the few at the expense of the many."[55] In response to their critics, fish and game commissions tried to minimize the elitist image of sports fishing. "It should be taken into consideration that the so called game fishes are the highest order of fishes, and that the love of angling is on the increase. The people from the farm, shop, store, factory, pulpit, studio, counting room, and court find a healthy relaxation from their cares in angling. The commercial fisherman and angler both have their rights which we are bound to respect. The whole people must be considered in the matter of propagating and planting fish in the waters of the state."[56] Indeed, the fish commissioners argued, commercial fishers and sports fishers had a mutuality of interests, and the commission should serve both groups. "All true sportsmen and all who make fishing a business should pull together, as their interests are identical. This commission is established to aid, and not retard, any efforts in the right direction, and will gladly cooperate with you all."[57]

Ideology of Conservation

Fish restoration supporters also articulated a vision of conservation that incorporated an ecological view of nature, not dissimilar to that articulated by George Perkins Marsh a half century earlier. They, like Marsh, believed that game conservation went hand in hand with nature's balance, which also benefited the farmer and citizen who was not a sportsman. In a language similar to Rachel Carson's a generation later, they warned that "should it ever come to pass that our land be void of bird and game life and our waters destitute of fish, the abomination of desolation would be upon us. With these extinct, the increase of noxious insects would be such as to materially interfere with the raising of agricultural products while the waters would so teem with animal culpae and larvae that it would not only be unfit for domestic use, but would be a serious menace to health. Nature too moves in a mysterious way, and we must preserve the natural balance or pay the penalty for our own short sightedness." These conservationists believed that failing to protect nature in all its complexity would bring ruin, which they defined, typically, in financial terms: "Nature [was] an ever willing lender, but one who on settling day demands a high rate of interest."[58]

The ideology of conservation that the fish and game commissions articulated argued for an interdependent natural world that protected human existence in the economy of nature. If the diversity of the natural world were maintained by the commissions, nature would provide humans with a place where one could find peace and tranquility. Protecting that diversity would also maintain the natural world for future generations. These twentieth-century commissions' argument for nature as a place of peace and tranquility reflected their urban and suburban perspective. Nature was no longer the location of food and sustenance. It was not a place needing to be tamed or husbanded. It was something to be visited. "All of you who love nature in any of her visible forms, and who on occasion at least . . . [find] rest and peace . . . near nature's heart, do not fail to often take [your] children with you on your fishing, or hunting trips and camping excursions."[59]

The fish and game commissions also linked their conception of conservation to the ideology of manhood, as had Marsh. And manhood and conservation were also linked to patriotism. The fish and game commissions reminded New Englanders that in appreciating nature, they were appreciating America. In teaching youth about nature, they were reinforcing national pride. "You can instill . . . patriotism into the child's mind by means of a woodland lesson or pointing out the beauty and freedom, peace and good will that abound in some shady nook. . . . The man who does not love the woods in which he gathers nuts and shoots squirrels

Twentieth-century sports hunters.

and birds, the lakes and streams from whose cool depths he has lured the finny occupant . . . may be fit for treason, stratagem and spoils, but is poor stuff to make a patriot of. Give the children a chance to become well acquainted with mother earth and some of the secrets hidden in her kindly old bosom."[60] This linking of conservation with manhood and patriotism was not unique to New England. Teddy Roosevelt made the same arguments in his defense of national parks and for the National Conservation Commission. Daniel Carter Beard picked up on this idea for his Sons of Daniel Boone and the Boy Scouts.[61]

Conservation and Americanism

By 1920, the argument linking conservation with manhood and nationalism had become a dominant part of the discourse of the fish and game commissions. With rather graphically phallic images, they argued "that a man is a better man if he longs to go afield with rod and gun and dog . . . and that . . . longing brings him into closer contact with the best most uplifting things in life." These conservationists believed not only that hunting and fishing made for better men but that "the real sportsmen of

America are our best citizens—clean of mind and body, resourceful, strong and courageous.... The love of nature—of clean, vigorous sport in the open—is the antidote for the softening, weakening influences of modern civilization." For patriotic manly virtue to flower, it needed a place to practice and hone the arts of hunting and fishing. To provide that place, the fish and game commissions had to "battle ... to recover the lost heritage which our ancestors wasted and failed to protect."[62]

The resources these twentieth-century reformers wanted to protect and "recover" were resources not for the poor, but for the sportsmen. To protect resources for sportsmen was also to protect America. The sportsman was "the best type of American citizen," who would save America with "man's most wholesome companions, animate and inanimate—the dog, the gun, and the rod.... From the patriotic view point, field sports furnish a practical antidote to Bolshevism on the one hand and on the other the experience thus gained best fits men for the raw material from which armies are made. Can we afford to sacrifice either of these national advantages?"[63] In an attempt to widen the appeal for conservation for sportsmen, the fish and game supporters also argued that hunting would be an antidote to radicalism if the working class took it up. "It is the workingman and his family who need most the call of the wild. When their eyes are opened to the mysterious mystic powers of nature, their gratitude will be expressed in terms of better citizenship."[64]

The link between conservation and Americanism and nationalism became entangled in nativism. In 1904, the Connecticut Commission of Fisheries and Game argued that special hunting licenses should be required of foreigners. "This is considered a practical solution of controlling the increasing numbers of foreigners who destroy birds and game despite laws and wardens."[65] By the second decade of the twentieth century, fish and game commissions were complaining that "many aliens, owing to ignorance, lack sympathy with conservation. A law forbidding unnaturalized foreigners from possessing firearms is in our opinion of great necessity."[66]

In a survey on the status of fish and game in Connecticut done for the fish and game commission, Leonard Samford, a naturalist and member of the commission, noted that the state was small and had a large foreign population. What the foreign population of the state had to do with fish and game may at first have eluded readers, except for the commission's assumption that the very fact of being foreign made foreigners a threat to the state's game reserves.[67] Not surprisingly, given the assumptions of the fish and game commissioners, a significant proportion of those convicted of violations of fish and game laws were foreigners.[68]

Although the new focus on sports hunting and fishing directed the attention of the fish and game commissions away from their traditional concerns of fish as food for the poor, the commissions did not completely abandon concern over water pollution. In a letter to the Connecticut Commissioner of Fisheries and Game, Charles Townsend, the director of the New York Aquarium, noted that stocking fish is difficult if the water is polluted. "The great evil with which practical fish culture in America has to contend at the present time is the contamination of public waters by sewage and the refuge of manufacturers."[69]

Townsend may still have been concerned about water pollution in all the region's waters, but other fish restoration reformers argued that the fish and game commissions should simply give up on the heavily industrialized and polluted waters and concentrate on game fish in more secluded lakes and streams. Hugh Smith of the Connecticut State Department of Commerce, in a letter to Frederic Walcott of the fish and game commission, proposed that "the Connecticut River is no longer suitable for any migratory fish because of the large amounts of trade waste discharged into it and because of the barriers below the sections to which such fish would have to go for spawning purposes." Smith suggested that the solution was for the commission to abandon shad and focus on stocking bass and speckled trout.[70] Although the fish and game commission was not willing to take up Smith's suggestion to abandon shad in the Connecticut River entirely, by the 1920s, sentiment had clearly moved to accepting that salmon would never be brought back and should be abandoned entirely as a project.[71]

Fish commissioner and naturalist Leonard Samford believed it was still possible to restore shad to the Connecticut, but like Smith, he also felt the commission should focus on protecting the habitat for sport fish, particularly trout.[72] Reflecting a disdain for the casual sportsman, Stamford argued that the sport fish situation in the state was "utterly bad ... because the state's ponds, rivers, and streams [were] easily accessible in autos. All ponds and other rivers are fished out in the early part of the season."[73]

Hunting and Fishing as Economic Goods

A focus on protecting habitat and breeding stock for sport fish and game meant that the economic justification for conservation, as opposed to the ideological one, shifted away from the benefits of cheap food and employment of commercial fishers to the economics of sport fishing and hunting. Seeing game conservation as a vital part of the region's economy had its roots in the late nineteenth century, although it was downplayed at the time. In 1895, the Massachusetts Fish and Game Commission reported that "hundreds of thousands of dollars are expended by our people in Maine and New Hampshire in pursuit of fish and game. Would it not be economical for the state to protect and increase our fish and game, so that a portion of this money be retained at home?"[74] Two years later, the Connecticut commission argued "that the good old state of Connecticut with its . . . sparking lakes, with its health giving breezes, needs but plenty of fish and game to make it still more attractive to summer and fall visitors of other states. These people spend money."[75]

Concern for maintaining game populations and a ready supply of fish for sportsmen required the fish and game commissions to focus more of their attention on game limits, hunting and fishing season restrictions, and programs for fish and game restoration. As more people took to the fields and streams to hunt and fish, pressure on the game populations also increased. "The natural passion of mankind for hunting makes the subject of game protection one of almost general interest." Facing that need, the commissions pushed for more and more regulation, to protect game. "The principle of game protection is, in abstract, popular today with the whole people, except those whose selfish interests is to profit by the rapid and unreasonable slaughter."[76] Increasingly, the fish and game commissions saw sports fishers and hunters as their constituency and worked to mobilize them in support of their activity, and they saw those who hunted and fished for food as having "selfish interests." After 1895, fish commissions spoke directly to or for "sportsmen." And it was "sportsmen," not sports men and women, to whom they spoke. "If American sportsmen wish that sport in the open with gun and rod shall sanely and sensibly be saved from extinction, . . . they need to secure it."[77]

As the focus of the commissions shifted toward sports fishing and hunting, more of their reports emphasized the values of good sportsmanship. In that discourse, the hunters and fishers were not likely to be the "workingman" whom the commission wanted to get out to the "wilds" or even the commercial fishers, but rather gentlemen. "All gentlemen sportsmen will respect the rights of owners who post their lands against hunting." Those who failed to comply with game restrictions were depicted as not sportsmen, but as "game-hog trespassers."[78]

Rights of Sportsmen over Property Rights

Although the initial constituencies for the early fish commissions had been the small commercial fishers and the marginal farmers and artisans who fished the rivers to supplement their diets, by the twentieth century the commissioners had come to see the rural marginal farmers and artisans as a problem for game conservation and as opponents to their new constituents, the gentlemen sportsmen. Frederic Walcott, in defending the right of sportsmen to cross over farmers' fields and fences, claimed that such action was an American right. "In free America our laws against trespass on fenced property are a howling farce.... They represent the fetish of 'personal liberty' brutally thrusting aside the most fundamental of all property rights [the right to hunt and take wild game as property]."[79] Walcott's defense of hunters crossing private property was not the defense of an antiproperty socialist, but the defense of an urban sportsman who wanted to be able to go out into the rural countryside and hunt to his pleasure, without being restricted by the local farmer. It also shows how far conservationism had moved from Theodore Lyman, who tried to protect the private property of his friend Samuel Tisdale so that Tisdale could stock his local pond without fear of the neighbors coming in and fishing, to Walcott, who ranted against "laws against trespass on fenced property." In many ways, this shift reflected the increasing urban isolation of the sports hunter and fisher in the twentieth century. One suspects that in the nineteenth century, sports hunters and fishers, like Lyman, would have had friends with private rural estates to which they could go to fish and hunt. By the twentieth century, people of the same social class needed the rural area of marginal farmers to find enough game for their sport.

In 1905, the Connecticut commission complained that "in some sections of the state [the poorer areas] there is a manifest unfriendliness to the laws for the protection of fish and game, so much so that it is impossible to secure conviction in some of the local courts."[80] Although the commission claimed it was laws for the protection of fish and game that people were violating, the laws that people in the rural poor areas were most likely to violate were those proscribing Sunday hunting and hunting without a license.[81]

Even with the increase in woodlands in the state, due to the decline of marginal farming over the second half of the nineteenth century, conservationists confronted an increasing human population and ever greater pressure on fish and game.[82] Despite attempts to clean up water pollution, the problem continued to plague those interested in fish restoration. If the solution to the problems of declining fish populations for Theodore Lyman and his generation was science and technology, the solution to the problem of pressure on wildlife populations for these twentieth-century reformers was "scientific business management," the mantra of the twentieth century. Not surprisingly, Walcott, the businessman, believed that "in a word, the most modern scientific business management is essential for success."[83] In their report to the state, the Connecticut Commission of Fish and Game argued that "the administration of a strictly efficient and up to date conservation policy has called for strenuous lines of management and exceptional executive ability on the part of the superintendent."[84] These reformers also believed in the ability of paid professionals to solve the problems of conservation. "The work of the warden should be regarded as a profession rather than a temporary job and the officer must do more than ordinary police duty to succeed."[85] Not only would "paid wardens ... promoted on the merit system" carry out the new scientific business management, but the wardens would be backed by "a trained, skilled secret-service force."[86] Who, one might wonder, would these "trained, skilled secret-service force" agents be used against? One suspects the targets of this force were those very marginal farmers whom the earlier commissions of inland fisheries were so interested in, and, of course, "the foreigners."[87]

Although the twentieth-century commissions on fish and game focused their attention on game management for sportspeople, the older issues of migratory fish and pollution never completely disappeared. In 1925, following the rapid decline in the numbers of shad the preceding few years, the State of Connecticut charged the state fisheries and game commissioners to study and report back on the condition of the fish in the Connecticut. Between 1870 and 1880, after the establishment of the initial hatcheries, "there was a great abundance of shad ... especially in the Connecticut itself." The catch fell off over the next two decades but began to increase again at the turn of the century. Beginning in 1907, shad catches dropped off to a third of what they had been earlier. Although there was limited improvement between 1917 and 1920, the catch started to fall again in the 1920s, reaching new lows in 1922 and 1923. Anxiety grew that indeed the shad were on their way to extinction in the Connecticut.[88]

The report of the investigation on the declining shad population raised concerns that had been around since the first commissioners of inland fisheries met at Theodore Lyman's Brookline home sixty years earlier: dams and overfishing. Reflecting a growth in science from the age of Lyman, the report also raised or understood in new ways the issue of pollution, which had been downplayed by Lyman. "Gradually sludge collected on river bottoms in the deeper places, these are just the locations which young shad haunt and where they find their food. Also in these deeper spots fertilized shad eggs ... tend to collect. ... Sludge and soft mud envelope the eggs and smother them before any hatch. Such material also depletes oxygen [which] ... cuts down the abundance of the small animal forms which young shad feed on and thus decreases their chances for successful growth. ... Sewage and other pollution ... cause serious limitations of the success of propagation of the species."[89]

Although the authors of the report were concerned about pollution, like Lyman sixty years earlier, they felt that shad could survive provided they avoided the heavily polluted areas and "followed the channels of better water." Yet despite conditions "comparable to ... a cesspool" at

Windsor Locks, Hartford south to Rocky Hill, and Middletown, the report's authors noted optimistically that pollution had not taken over the whole river.[90]

Reflecting, perhaps, the shifting focus of fish and game commissions in the twentieth century, Walcott, the main author of the report, concluded with the rather ambiguous message that "for these various reasons [dams, pollution, and overfishing] and possibly others, the decline of shad abundance might well be expected. It is, of course, no more surprising than the depletion of salmon or any other species of wild life sought by man."[91] Ironically, after sixty years of struggle to restore fish to the region's rivers, Walcott seems to accept the view of Jerome Smith that the decline of nature's abundance, at least in the major rivers, was an inevitable consequence of civilization.

If the progeny of the Theodore Lymans and of Judge Henry Bellows were more concerned with sport fish and game than a restoration of the earlier bounty of the region's waters for fish as food for the citizens, they also shared with Lyman a belief in the necessity of conserving resources, the ability of science to overcome difficulties, and the importance of the state to act to protect resources against the greed and self-interest (or ignorance) of the individual. In the case of the twentieth-century conservationists, the science was the science of scientific management, the resources were game habitat, and the power of the state was wardens enforcing hunting and fishing restrictions. And the 1920s conservationists proved no more willing to challenge manufacturers than had their nineteenth-century counterparts. Avoidance seemed the more prudent route. To be fair to Walcott and his compatriots, they partly failed to challenge manufacturers because, although they saw the pollution in the rivers and streams, that pollution seemed an inevitable consequence of time itself. They, like Lyman, came from a social class that was intimately tied to the world of the manufacturer and manufacturing. They were also gentlemen sportsmen who appreciated nature and nature's beauty. They believed that they and their fellow sportsmen were the saviors of the Western world. They wanted to protect nature so that they could appreciate it. At the same time, they did not want to pursue a course that would threaten their source of privilege. For Lyman and his generation, science, technology, and an active state could be mustered to protect natural resources in such a fashion as to allow society to have both industrial development—moderately restrained—and nature's bounty. For Senator Frederic Walcott and his generation, scientific management, gentlemanly sportsmanship, moderation in enjoyment of those resources, and the effective professional police power of the state over the poor and foreign born would allow for both nature's bounty and industrial development. In some ways, even this summary is unfair to Walcott and his fellows, for by 1925, the very issue of confronting the manufacturers never even crossed their minds.

These conservationists did leave us, though, with a heritage we should not forget. For while Henry Ford was turning out automobiles and waxing nostalgic over the lost world of rural America, and while Samuel Insull was putting together the great utility trusts, members of the various fish and game commissions volunteered their time and energy to promote the appreciation and protection of the quickly disappearing fauna of New England.[92]

11

New England, the Nation, and Us

Sylvester Judd died in 1860 at seventy years of age, just before he completed his history of Hadley, Massachusetts. Judd had lived through the transformation of his community from a small rural village where people fished massive runs of salmon and shad each spring, to an industrial center seated on the banks of a polluted river. One of his motivations for writing his history was to capture that fast-disappearing older world. It was a world where lawyers, shopkeepers, journalists, and farmers (and Judd had been three of those four) knew how to cut timber, kill and clean a turkey, catch fish, butcher a pig, tend a garden, work an orchard, and make cider. By the time of Judd's death, wood was sold already cut into cordwood or milled to clapboards, meat was butchered at the abattoir, and cloth was woven in mills. It was a world that could not be brought back through his history, but one that Judd hoped through his history might at least be remembered.

On July 24, 1882, when Theodore Lyman went before the people of his district to run for Congress, he reminded them that when he was a boy (then, Judd was in his fifties), the region's industry had already begun to grow, although many of the state's residents were still rural farmers. Yet by the time Lyman ran for Congress, a majority of the people of Massachusetts found their homes and their jobs in towns and cities. The world that Judd saw fading in the 1850s was indeed a thing of memory for some of those listening to Lyman in 1882. New England of the 1880s was a place, as Lyman noted, of "manufacturing towns with . . . sickly smells." Yet without this progress, according to him, New England would have remained a place of "a few grist-mills here and there and houses whose occupants raised such crops as they could from the scanty soil."[1] If, by 1882, the people of New England had lost their more direct contact with the resources of nature, for many, the romanticized memory of that intimacy lingered on.

Some New Englanders, like Theodore Lyman, saw their interests and the interests of the region tied together in the success of the industrial becoming. "I am not an old man, and yet when I think of the state of our manufactories in my boyhood and when I look at them today, I can hardly believe my eyes such is the enormous growth in variety, in technical skill, and in inventive power."[2] But as Lyman reminded his audience, his family was "connected . . . with the manufacturing interest."[3] Some, like the radical reformer Henry Bowditch, were not so much hostile to industrialization as opposed to the destructive consequences engendered by it. And they were willing to fight against the manufacturers to mitigate those consequences. Others, like the ever litigious Royal Call or like Samuel Ely, who tried to run off the corporation lawyers with his gun, were more ambivalent or even hostile to the transformation itself. Even those who supported the transformation, like the avid naturalist, scientist, and investor in manufacturing Theodore Lyman, believed it should not be left to its own direction but needed to be controlled and moderated.

All these responses grew out of a time when people remembered an either real or imaginary natural world not as a place to sojourn, but as a way of life. Their responses came about when manufacturers were creating and building a world of industry and taking control and ownership of nature and nature's resources in new and ever more dramatic forms. Urban industrial New England was not a given, but something being made. Although Lyman so heartily but ambivalently embraced the progress that industrial manufacturing represented in the middle of the nineteenth century, other New Englanders were not fully in the grip of the ideology of progress, and the ideas that progress and pollution were inevitable. In the process of environmental change, their voices rose to challenge how that transformation proceeded, and those voices, indeed, did affect the nature of that transformation.

The reformers may not have directed it quite the way they wanted, but they did change the way the natural world was transformed. In doing so, they also left a legacy to future generations. The urban world became safer because of the work of state medicine. Typhoid, cholera, and dysentery ceased to be the ravaging killers of urban dwellers they had been at the turn of the century. The mortality rate for males in Massachusetts, for example, fell from an average of over twenty per thousand in the population in the 1870s to the low teens per thousand by the 1920s.[4] Yet as public health crusader Edgar Sydenstricker noted in 1933, looking back over the last half century of public health work to improve the environment, "a reasonably healthy environment . . . has not yet been attained."[5]

Humans have long been well aware of the changes they have created. When they burned forests to clear ground for camps or to scare up game;

when they piled up stones and boulders to create dams to trap fish or flood fields; when they cut trees and cultivated fields, they knew they were changing the natural world. But as far as we can tell, these acts transformed the natural world so slowly or so slightly that a deep sense of stability existed in the memory of the transformers. At the end of the eighteenth century, the pace of change itself began to speed up. And it sped up so fast that during the nineteenth century, it was changing faster than generations themselves were changing. In that century, humans could remember a world radically different from the one in which they lived. That realization brought with it the consciousness that people could control the transformation itself. They were entitled not only to a life in the world but to the pursuit of happiness.

In New England, the nineteenth century's radical transformation of the natural world engendered by industrialization and urbanization gave rise to protests over what was happening to the environment, from the fouling of pure water to the depletion of fish and game. The protesters understood that the natural world was being transformed by the hand of man, or more particularly, by an instrument in the hand of man. Their protest set in motion a reform movement.

The change occurring about them was also altering the world in a way that forced reformers to explore new ways of responding. Long before industrialization, New Englanders were accustomed to coming together as individuals and as members of a community to air grievances, adjudicate conflict, and facilitate cooperation. When the Shaysites met in Hatfield in 1786 to protest taxes and foreclosures, they were participating in a political process that had its roots in both the Revolutionary experience and in a tradition that went back to the times of the ancient charters. The county convention, as well as the town meeting, the general court, and the court of common pleas were embodiments of the place of collective protest or adjudication. Of course, not everyone was equal before the court, but neither were they (if they were free males) unrecognized. Most free males got their hearing, and if the mill owner was favored (as in the Mill Acts of 1796) over the meadow owner or fisher, it was presumed that favoring the mill owner was for the public good and that the farmer would benefit too.

The Mill Acts were passed when New England was a rural society of farmers, tradesmen, and artisans. But the world of the small farmers and local sawmills and gristmills that gave rise to the original Mill Acts was gone by the second half of the nineteenth century, as both Lyman and Judd realized. Over the first half of the century, court cases and statutes had created and empowered a new entity, one that did not come to the court of common pleas or even the general court as an individual and a member of the community, one that had the rights of an individual but the immortality of the gods. The new corporations altered the nature of the political world, as they changed the natural world. In changing both the natural world and the political world, they also altered the nature of New Englanders' understanding of the state. The change rendered by the corporations and their industrial processes undermined the world of the individual as a member of the community. That transformation also created a context and a situation in which an individual at the court of common pleas or as a member of a local community could no longer deal with the larger environmental disruption.

The new corporations were powerful entities that dwarfed the resources and abilities of the individual. And the changes that these large corporations as a group brought about in the environment, from the massing of people into industrial mill towns to the dumping of huge amounts of wastes into the water, were beyond the ability of the individual to affect. In a world where river water ran through several states and picked up pollution from dozens of mills—paper mills, woolen mills, cotton mills, and machine shops—and from just as many towns and cities, against whom could one take tort action? How could one individual address such a mess? Once-plentiful fish now disappeared because their spawning grounds were fouled and muddied due to the cutting of forests for lumber for tenements and homes in towns hundreds of miles away. Dams blocked the fish migrations, pollution killed the fish as they swam, and fishers overcaught them. Who did one sue? What action could an individual take? When fewer and fewer individuals owned property but lived in a world increasingly compromised, what protection did the "sacred right of private property" provide?

The mills and factories that brought the region wealth also brought the region pollution, enough to dwarf the problems of environmental change of earlier generations of New Englanders. Water that had previously run clear and clean and abundant with fish now was "undrinkable" and "poisoned." In such a world, who did one take to the court of common pleas? Finding the agent responsible for the massive damage occurring throughout the land became an impossibility. The change was too great and too amorphous, and it stretched over too long a period of time.[6] Court decisions over the length of the nineteenth century had distanced corporations from liability.[7] Many of the changes that occurred in the natural world were not caused by a single corporation but by the larger transformation itself.

Because of court decisions and legislative action that limited nuisance suits, and because of the increasingly complex nature of the problem, traditional modes of redress seemed to fail. This failure led many New England reformers not only to challenge the practices that led to this environmental devastation but also to push for a more aggressive, proactive state. Pollution reformers articulated a new role for the state in dealing with the impact of pollution on the region's environment. The Massachusetts State Board of Health noted in 1874 that "it is certainly the duty of the government to protect the weak from oppression of the strong. . . . It is this last class that suffers most from . . . unwholesome surroundings and other unsanitary conditions which can only be controlled or suppressed by official effort." The board continued that "it is only a question of time how long it will be before each state must provide some official means to also protect the public at large."[8] In 1882, the New Hampshire Board of Health reminded its citizens that "no individual or association can undertake and carry alone such a labor [the protection of the environment] as well as the state."[9]

In that context of change, people looked to the state as an agent that could act to protect the common good. In pushing the state to protect the common good, these first-generation reformers, despite their shortcomings, managed to raise basic issues in a fashion that makes them pioneers of modern environmentalism. In the nineteenth century, when New Englanders saw their countryside transformed from "the most beautiful" to "open sewers," they laid the foundation for a more aggressive state.[10]

New England's Past as Our Present

These New England reformers demand our attention not only because they pioneered in environmental concerns, but because their concerns are also ours. They were not trying to protect a relatively unsettled wilderness. Instead, they struggled with the questions of how to maintain and preserve a livable space in an environment already compromised by urbanization and industrialization. It was not so much the rural cabin in the wilderness that concerned them as it was the tenement in the city. And it was less the aesthetic musing of a sojourner into the wilds of Yellowstone than the need of the "weary mother [who] takes her infant from some crowded tenement-house" for a clean park and fresh water running through it that concerned them. In their actions, we might also find lessons for our struggle to find solutions to the problems of making our environment a livable place.[11]

Although the waters of New England were in many ways no cleaner in the 1950s than they were in the 1870s, thanks to the work of these early reformers they were less dangerous.[12] Although today we might decry the nineteenth-century habit of culverting and covering over streams rather than protecting them from degradation, at least the smell of human waste decreased in many urban areas. Continued agitation against the dumping of raw sewage into waterways did encourage some communities to build sewage treatment plants. Salmon may not have returned to the Connecticut, but by the end of the nineteenth century, New Englanders could claim to have at least the presence of both fish and game, where a generation earlier, it had been assumed that such resources would forever be extinguished.[13]

These New Englanders confronted hard choices. Industry that brought wealth and prosperity, at least for many of the region's inhabitants, also brought real costs in terms of added industrial wastes and pollution and reduced fish in the rivers. When the reformers attempted to deal with the problems of environmental decline, the manufacturing interests balked because they saw these reforms as a threat to their profits. The corporations lobbied against reform legislation, and they tied up reforms in the courts. They threatened to move their businesses elsewhere and leave behind unemployed New Englanders.

The manufacturers were no mean opponents. Since the 1830s, the courts were inclined to favor the manufacturing interests over those of the farmer or fisher. Wealthy, powerful owners of the large manufacturing corporations exercised significant influence in the state legislatures and often had a friend in the governor's office. This opposition at times defeated the reformers outright or forced them to compromise. But the reformers were not completely routed from the field. If they lost or were compromised, they also came back. To focus on their defeats or compromises is to miss their greatest accomplishment.

In a rapidly expanding and changing economic world in which industrialists encouraged seeing New England only as a place of jobs—a place of work and a place to make money—the reformers refused to accept this narrow vision of human existence. They continually refocused the discussion around society as a place to live. When challenged by the industrialists that society needed to protect its industry, they reminded New Englanders that "the first and largest interest of the State lies in this great agency of human power—the health of the people," and that "the protection of the health of the people . . . should be looked upon with as much consideration as given to the care of property or the fostering of productive industries."[14] They argued that the state also had the responsibility to protect the environment not only for the living but for future generations, "my children's children," as Edward Everett Hale stated.[15]

The reformers also left behind a legacy of contested power. Although the industrialists time and again beat back the reformers in their attempts to pass legislation prohibiting the dumping of pollutants into the region's waters, the reformers managed to insert the issue of public power over the environment into political debate. As James Olcott reminded the farmers of Connecticut in 1886, although the industrialists were powerful interests opposed to "anti-stream pollution," "the mightiest vested interest in the land . . . could not prevail against [the agitation of the common people]."[16]

These New Englanders also deserve our attention because their pioneering conservation established the model for other states. Shortly after Massachusetts set up its board of health, other states looked to it for a model when establishing their boards.[17] The New England pioneers also took their zeal for reform to the national level. In 1871, within a few years of the creation of the New England commissions on inland fisheries, Congress passed a resolution asking the president to appoint a commissioner of fish and fisheries. Professor Spencer F. Baird of the Smithsonian Institution, chosen to be the first commissioner, was a close friend of Lyman and had a home and research laboratory in Woods Hole, Massachusetts. Lyman acted as an advisor to the National Fish and Fisheries Commission.[18] Bowditch gave the centennial address at the International Congress of Medicine in 1876 on the subject "State Medicine and Public Hygiene in America." When the nation established a national board of health, he was appointed by the president as one of its first members.[19] Henry Walcott not only became chair of the Massachusetts State Board of Health in 1886 but was also elected president of the American Public Health Association.

These early reformers articulated a vision of state power that was national in scope. When a Dr. Bell wrote to Henry Bowditch asking how the nation could best protect human health, Bowditch responded with a vision of state power: "the time will come (if it be not already at least partially arrived . . .) when the whole people deem the health of the citizens [to be] the care of government."[20]

By the early years of the twentieth century, the idea of the activist state had become associated with the Progressives. Historians such as Samuel Hays and Robert Weibe have pointed out how concern over the problems of urbanization and industrialization encouraged the Progressives to push for more state action and regulation. Historians following the work of Mary and Oscar Handlin have also been aware since 1960 that the laissez faire state was hardly a sleeping giant in the nineteenth century.[21] Although state action to promote economic growth, establish the framework for safety, and mediate among its citizens had long been accepted, the Progressives moved the framework and mediation aspect of state action into the foreground and highlighted it as a positive function that the state should play. The Progressives articulated a vision of a more proactive state. In so doing, they followed upon a legacy of reform articulated almost a half a century earlier in New England, where residents confronted the problems of urbanization and industrialization not only as those problems manifested themselves in riots, strikes, and poverty, but also as they manifested themselves in a degraded environment. A full generation before Theodore Roosevelt and Gifford Pinchot argued for professionalizing the management of resources and bringing science and technology to bear on environmental problems in order to have both economic development and resource husbandry, New England reformers were struggling over those very same issues and articulating positions later identified with the Progressives.

Ironically, what the Progressives embraced most aggressively from the earlier New England experience was the optimism of the moderate reformers like Lyman and Mills. It was an optimism that believed that scientific and technological expertise in the hands of professionals supported by the state could transcend the conflict of class and economic interest. That optimism was rooted more in faith than in history. For although a cursory view of the New England experience and certainly the rhetoric coming from the reformers would give credence to the optimism, a more penetrating look at the history of these pioneer reformers might have made the Progressives more sanguine.

In 1938, the Works Progress Administration did a study of the Connecticut River and found it as polluted as when the reformers first began agitating "to cleanse" the water of pollution, and for the same reasons. "Because of the heavy concentration of both population and industry, pollution on the Connecticut River, especially in the lower reaches near Holyoke, Springfield, West Springfield, and Chicopee has at times been so heavy as to cause the Department of Public Health to restrict the use of the river for bathing."[22] In 1950, the Federal Security Agency of the U. S. Public Health Service again looked at pollution in the Connecticut, and the river again came up short.[23]

If Henry Bowditch had lived long enough to see these reports reflecting such a dismal record of accomplishment, he might have been as disgusted as he was in 1881. Disgusted Bowditch might be, but discouraged, I doubt. For Bowditch left us with not only a legacy of concern for the "rights of clean air, clean water, and clean soil" but also the legacy of struggle. He understood, as James Olcott said, that the citizens had to "agitate, agitate" to "cleanse" the region of the "social evil" of pollution.

It is a complex and ambiguous history that this first generation of environmental reformers has left us. But in their struggle to establish the right of "all citizens . . . to the enjoyment of pure and uncontaminated air, and water, and soil, [and] that this right should be regarded as belonging to the whole community, and no one should be allowed to trespass upon [it] by his carelessness or his avarice," they have given us an unambiguous legacy and one we need to remember and honor.

Epilogue

Residents along the Connecticut River seventy years ago would not recognize the river today. On a typical Sunday afternoon, one can find people bathing, fishing, or boating in a relatively clean waterway. Since the passage of the Clean Waters Act of 1967 and the spending of over a billion dollars to build water treatment facilities along the river, the Connecticut's water quality has improved enough to move the river from class C (no fishing or swimming) to class B (both are allowed). Riverfront parks are springing up in cities that had long seen the river only as a flood threat or a convenient dump.

The change of the Connecticut was slow in coming. Initially, most of the concern shown for the river in the period after 1927 was related to flooding. After the 1927 flood, leaders throughout New England realized the region faced a crisis. The Rivers and Harbors Act, which brought federal relief to the area, also encouraged regional flood planning and the beginning of major dam building for flood control and hydroelectric power. In 1936, the New England Regional Planning Commission was established, with a water resources committee. This was followed by a series of federal acts to control flooding. With these acts, the upper reaches of the river were remade by the human hand. Valleys became lakes, and electric power flowed out of the northern valley to the cities of the south, as lumber once had.[24] But although the river was controlled as never before, in the decades before 1970, progress at cleaning the river was slow. Yet the story of the Connecticut River's rebound from a class C to a class B river did begin long before the 1967 Clean Water Act. It began with the work of the first generation of environmental reformers, for they were the pioneers.

The nineteenth-century reformers who could remember an environment significantly less degraded had an especially valuable vantage point from which to develop their critique of the new urban industrial world. In the communities of New England in the second half of the nineteenth century, they began to argue that air, water, and soil were being despoiled. Theirs was a world that should resonate with those of us today who live in, or at the edge of, an urban wilderness rather than an old-growth forest. And just as nineteenth-century New England reformers confronted resistance, so too do today's environmentalists. The problem that confronted both moderate reformers like Lyman and radicals like Bowditch was the conflict between development and the environment. Nineteenth-century manufacturers, like their counterparts today, were resistant to any reforms that threatened their prerogatives, particularly what they believed was their right to maximize profits and minimize costs. In the nineteenth century, as today, manufacturers hoped to minimize their costs by externalizing them. Indeed, the very process of manufacturing generated toxic and complex wastes. Not knowing how to deal with these inconvenient, noxious, if not toxic, byproducts, manufacturers simply dumped them into nature's environmental lap. Any attempt to have the manufacturers take them back or even to limit the manufacturers' dumping was met with stiff resistance.

Faced with the reformers' campaign to ameliorate environmental degradation, the opponents of reform defended economic development as a natural good and its attendant pollution as a natural given, a byproduct of a good that could not be helped and was as inevitable as progress itself. The opponents of reform also buttressed their arguments with claims of the sanctity of their particular forms of property. The reformers called for a restoration of natural flora and fauna abundance. The manufacturers were instrumental in increasing a new form of abundance and wealth. While the reformers claimed to be interested in the poor and in cheap fish for food or in clean air, water, and soil, the manufacturers claimed to be responsible for creating jobs, industry, and prosperity. It was a prosperity skewed toward one group over another, to be sure, but as Lyman reminded New Englanders, without manufacturing, the region's poor soil would maintain only a meager existence.

Lyman drew New England's options as a choice between development and manufacturing on the one side, and poverty on the other. Henry David Thoreau rejected that view, defining development and manufacturing as a kind of poverty, impoverishing the land and waters of the region and the soul of the developer. If Thoreau rejected the development and manufacturing option, Lyman could not even consider Thoreau's antimodernist, simple-subsistence alternative. Yet Lyman was concerned about the impact of manufacturing and development on the natural world. For Lyman, progress could address the environmental problems inherent in development. Science and technology would bring new means of restoring nature's abundance without challenging the interests of forward-thinking manufacturers.

These issues speak to us today. The voices of Jerome Smith and Charles Donnelly, who claimed that we should not meddle in the affairs of the economy for the sake of nature or that there were always more resources farther afield, are echoed today in public debates, newspaper articles, and the halls of Congress. The nineteenth-century manufacturers' warnings that pollution reform or fish protection would hurt their profits and force them to close down and leave villages without jobs are warnings we still hear. The hopeful echoes of Lyman and the Connecticut State Board of Health that science and technology will provide economic development without natural resource degradation are also in the air. Among some environmentalists, Henry David Thoreau's rejection of progress and development resonates powerfully.

Today, as we hear the calls against environmental regulation, we should not forget that even before the age of toxic carbon-based pesticides and herbicides that Rachael Carson warned us about almost forty years ago, a silent spring was slowly encroaching on the New England countryside. Waters once abundant in salmon, shad, and herring were void or nearly void of them. And clear, crisp New England air was increasingly fouled by smoke and noxious odors. In the face of that degradation, New Englanders from a variety of backgrounds and perspectives took action.

Many of this first generation of New England environmentalists believed along with Lyman and Mills that conflict between development and the environment could be avoided with better science and better technology. It is a comforting view in that it holds out a promise of no hard choices, of a win-win game. Past history indicates it is also a chimera. Environmental protection, like manufacturing development, costs. Who bears those costs, Bowditch would argue, is a political question. For Bowditch, the citizens of New England had a right to clean air, clean water, and clean soil as fundamental as (and tied up in) their rights to life, liberty, and the pursuit of happiness. If a clean environment was a right, the costs of maintaining it or cleaning it up should rest on those that threaten or destroy it.

These early environmental prophets not only issued warning about environmental degradation, they also provided us in their successes and failures with lessons on how to approach those problems. These lessons we need to remember, because in many ways the problems set to us today were first set to them some 150 years ago.

Notes

CHAPTER 1

1. Chet Raymo and Maureen E. Raymo, *Written in Stone: A Geological History of the Northeastern United States* (Old Saybrook, 1989), 99–103.
2. Timothy Dwight, *Travels Through New England and New York* 4 vols., ed. Barbara Soloman and Patricia King (Cambridge, 1969), 224.
3. Raymo and Raymo, *Written in Stone*, 103. See William J. Miller, *The Geological History of the Connecticut Valley of Massachusetts* (Hampshire, 1942); Thomas Reed Lewis, "From Suffield to Saybrook: An Historical Geography of the Connecticut River Valley in Connecticut before 1800" (Rutgers University diss., 1978), 10, 11; and George W. Bain and Howard A. Meyerhoff, *The Flow of Time in the Connecticut Valley: Geological Imprints* (Springfield, 1963), 1–4.
4. Bain and Meyerhoff, *The Flow of Time*, 5, 8–10.
5. Edwin M. Bacon, *The Connecticut River and the Valley of the Connecticut: Three Hundred and Fifty Miles from Mountain to Sea, Historical and Descriptive* (New York, 1906), 351–352.
6. On the west side of the river, the principal tributaries are the Indian, Halls, Passumpsic, White, Ottaugueecheee, West, Deerfield, Agawam, and Farmington. On the east, they are the Upper Ammonoosuc, Lower Ammonoosuc, Ashuelot, Millers, Chicopee, Scantic, and Salmon.
7. Dwight, *Travels*, 75; Dwight, *Travels*, 2:58. Judd Papers, "Miscellaneous, Vol. 9, 74," 355; Dwight *Travels*, 4:332 Vol. IV, also Dwight *Travels*, 1:141, "Letter XV," 141. Theodore Dwight, *Notes of a Northern Traveler* (Hartford, 1831), 144, 151.
8. Dwight, *Travels*, 2:94.
9. Dwight, *Travels*, 1:35. Overhunting and the clearing of forests reduced the number of deer, turkeys, and partridges. Overhunting took its toll on wild ducks and wild geese. Sylvester Judd, *History of Hadley: Including History of Hatfield, South Hadley, Amherst, and Granby* (Springfield, 1905), 351.
10. The farmers of the valley were not the first inhabitants to give form and shape to the valley's landscape. See William Cronon, *Changes in the Land: Indians, Colonists, and the Ecology of New England* (New York, 1983), and Carolyn

Merchant, *Ecological Revolutions: Nature, Gender, and Science in New England* (Chapel Hill, 1989).

11. Merchant, *Ecological Revolutions*; Stephan Innes, *Labor in a New Land: Economy and Society in Seventeenth-Century Springfield* (Princeton, 1983), 5; Lewis, "From Suffield to Saybrook," 37–40; Dwight *Travels*, 1:98, 99; Christopher Clark, *The Roots of Rural Capitalism: Western Massachusetts, 1780–1860* (Ithaca, 1990), 22–23. Dwight *Travels*, 1:19. Margaret Pabst, *Agricultural Trends in the Connecticut Valley Region of Massachusetts, 1800–1900* (Northampton, 1941), 2; Lewis, "From Suffield to Saybrook," 19–22; Dwight 1:19. David Szatmary, *Shays' Rebellion: The Making of an Agrarian Insurrection* (Amherst, 1980); Robert Taylor, *Western Massachusetts in the Revolution* (Providence, 1954); and Will L. Clark, ed., *Western Massachusetts: A History 1636–1925*, vol. 1 (New York, 1926).

12. See Clark, *Western Massachusetts*, vii, 9; Clark, *Roots*, 8; Merchant, *Ecological Revolutions*, 149–152, and Pabst, *Agricultural Trends*, 18, for a discussion of populations of hill towns and valley towns. See also Robert A. Gross, "Culture and Cultivation: Agriculture and Society in Thoreau's Concord," *Journal of American History*, 69 (June 1982): 42–61; Harold Fisher Wilson, *The Hill Country of Northern New England: Its Social and Economic History, 1790–1930* (New York, 1967), and Richard Judd, *Common Lands, Common People: The Origins of Conservation in Northern New England* (Cambridge, MA, 1997) and Stephan Inn.

13. Lewis, "From Suffield to Saybrook," 226–230; Merchant, *Ecological Revolutions*, 112–197. Even as late as 1818, a "valuable working farm" was offered for sale in the upper valley with only two-thirds "under improvement." *Vermont Republican and American Yeoman*, Apr. 13, 1818.

14. Dwight, *Travels*, 2:322.

15. Ibid., 321, 322. As one valley farmer stated in 1786, "nothing to wear, eat, or drink was purchased, as my farm provided all." *Massachusetts Centennial*, June 24, 1786, quoted in Szatmary, *Shays' Rebellion*, 6. See also Pabst, *Agricultural Trends*, 26.

16. Dwight, *Travels*, 1:76. Into the late eighteenth century, Connecticut River Valley farmers planted maize over flax, even though flax commanded a higher market value, because flax was more labor intensive. Dwight, *Travels*, 2:321, 322, 83. See Christopher Clark, 28 and also Alan Taylor, "Unnatural Inequalities: Social and Environmental Histories," *Environmental History* 1:4 (Oct. 1996):13–15.

17. In Hadley at the end of the eighteenth century, there was pasture land for only 205 cows, yet the town had 468 cows, oxen, and horses, and 603 sheep. The surplus animals were let out to fend for themselves in the woods for the summer months. Judd Papers, "Miscellaneous, Vol. 13, 109," for other examples of animals foraging see also 352; "Miscellaneous Vol. 9, 377," 493. Judd notes that "pasturing domestic animals in the woods" was common in the seventeenth and eighteenth centuries. "The common fields and private lots required strong barriers to protect them against restless, rambling animals." Judd, *Hadley*, 103. Elihu Warner claimed that in Hadley the cows went

into the woods throughout the eighteenth century. Judd Papers, "Miscellaneous, Vol. 18, 218," 351. Early nineteenth-century farmers in the upper valley were still registering with the town the distinguishing marks (usually a clipping or cutting of the ear) they made on their animals before setting them free. Lyman Hayes, *History of the Town of Rockingham, Vermont, Including the Villages of Bellows Falls, Saxtons River, Rockingham, Cambridgeport, and Bartonsville, 1753–1907* (Bellows Falls, 1907), 98.

18. Judd Papers, "Miscellaneous, Vol. 12, 151," 119. See Steven Hahn, *Roots of Southern Populism: Yeoman Farmers and the Transformation of the Georgia Upcountry, 1850–1890* (Oxford, 1983), for a discussion of conflicts over common grazing and fencing.

19. See *Massachusetts Record of the Supreme Judicial Court* for the years 1804–1840.

20. Gross, "Culture and Cultivation."

21. Ibid.; Judd Papers, "Miscellaneous, Vol. 18, 387," 245, 246. Dwight, *Travels*, 4:249; Judd Papers, "Hadley, Vol. 3, 137," 347. Fresh meat of all kinds was typical dinner fare in the fall and winter, while in the summer the family ate salted beef and pork, sausages, and dried beef. Dwight, *Travels*, 4:249; Judd Papers, "Miscellaneous, Vol. 9, 254, 256," 244.

22. Judd Papers, "Miscellaneous, Vol. 15, 440," 235.

23. Isaac Weld Jr., *Travels Through North America and the Provinces of Upper and Lower Canada, 1795, 1796, 1797*, 4th ed. (London, 1807), 44.

24. Judd Papers, "Hadley, Vol. 7, 102."

25. See Judd, *Common Lands, Common People*.

26. Dwight, *Travels*, 2:325. "Fishermen are prone to take little care of their earnings." Dwight, *Travels*, 1:165. See Harry L. Watson "The Common Rights of Mankind: Shad and Commerce in the Early Republican South," *Journal of American History* 83, 1 (June 1996), 13–43, for a discussion of shad fishing in the South, its importance to subsistence farmers, and the view of the gentry that it encouraged indolence.

27. Judd Papers, "Hadley," Vol. 7, 102"; Dwight, *Travels*, 1:35. See John Cumbler, "The Early Making of an Environmental Consciousness," *Environmental History Review* 15 4 (Winter 1991): 75.

28. Alfred Booth interviewed old-timers who recounted stories about the abundance of shad in the early nineteenth century and even salmon in the eighteenth, and their importance for the local community as both a source of food and extra income. Connecticut Valley Historical Society, *Papers and Proceedings of the Connecticut Valley Historical Society, 1876–1881*, Vol. 1, Oct. 2, 1876 (Springfield, 1881), 16, 17, 18.

29. JCS, "Joint Special Committee Report," MA Sen., no. 183 (Boston, Apr. 1865), 5.

30. *Commonwealth v. Jonathan Knowlton* (June 1807), MA Reports, Tyng 2:529.

31. *Edward Vinton v. Jonas Welsh* (Oct. 1829), MA Reports, Pickering 9:87.

32. Lyman S. Hayes, *The Connecticut River Valley in Southern Vermont and New Hampshire: Historical Sketches* (Rutland, 1929), 174; Richard

Wilkie and Jack Tager, eds. *Historical Atlas of Massachusetts* (Amherst, 1991), 20, 25. Judd, *Hadley*, 307; Dwight, *Travels*, 1:165, 2:213, 224. See also Judd Papers, "Hadley, Vol. 7, 108," 63, 81. Lamprey eels came up "in great numbers" and "were caught by the light of torches sometimes several hundred a night." Judd, *Hadley*, 309, note. Daniel Lombard, Haravey Sanderson, and Festus Stebbins remembered that in their youth, shad "was considered poor men's food." And although shad were cheap, often less than two cents apiece, they were plentiful and an important addition to the diet. Connecticut Valley Historical Society, *Papers and Proceedings*, 1:16, 18.

33. Judd Papers, "Hadley, Vol. 7, 108."
34. Judd, *Hadley*, 309.
35. Judd Papers, "Hadley, Vol. 7, 'Shad.'" See also Judd, *Hadley*, 308.
36. Judd, *Hadley*, 305.
37. Ibid., 309. Matthew Patten, *The Diary of Matthew Patten of Bedford, New Hampshire* (Concord, 1903).
38. See Clark, *Roots*, 30–35.
39. See Allen Kulikoff, *The Origins of American Capitalism* (Charlottesville, 1992), for a discussion of the debate about the role of exchange and the question of market orientation.
40. Margaret Martin, "Merchants and Trade of the Connecticut River Valley, 1750–1820," *Smith College Studies in History*, vol. 24. nos.1–4 (Northhampton, Oct. 1938–July 1939), 5. See also Pabst, *Agricultural Trends*, 12–14. According to Pabst, the store owners carried these transactions on their books in dollar (earlier in pounds sterling) figures.
41. Like most country storekeepers in the late eighteenth and early nineteenth centuries, Richard Bigelow of Brattleboro offered to exchange "English and Indian goods" for rye. *Brattleborough Independent Freeholder and Republican Journal*, Dec. 8, 1808. Judd Papers, "Hadley, Vol. 3, 137," 346, 347, 350, 351; "Miscellaneous 15, 356," 341; "Northampton, Massachusetts, Prices and Account Books 113," 338, 337.
42. See notices in the *Hampshire Gazette* in the 1780s.
43. For a discussion of the abandonment of traditional agriculture for market-oriented agriculture, see Clark, *Roots*, and for a discussion of demographic pressure on traditional agriculture and the shift to improvementist agriculture, see Merchant, *Ecological Revolutions*. See also Pabst, *Agricultural Trends*. Although the evidence in New England is mixed, some economic historians have argued that pressure for consumer goods originated with industrious farmers. See Jan De Vries, "The Industrial Revolution and the Industrious Revolution," *Journal of Economic History* 54 (June 1994):249–270; and Maxine Berg, "Women's Work, Mechanization, and the Early Phases of Industrialization in England," in Patrick Joyce, ed., *The Historical Meaning of Work* (Cambridge, 1987), 69–76.
44. *Hampshire Gazette*, Nov. 10, 1789.
45. Szatmary, *Shays Rebellion*, 16, 17.
46. *Hampshire Gazette*, Oct. 18, 1786.
47. Ibid., Dec. 13, 1786.

48. Ibid., Dec. 6, 1786. In the 1780s, 72 percent of Oliver Dickinson's customers at his store in Amherst paid in goods and labor. Szatmary, *Shay's Rebellion*, 31.

49. Dan Butler, who also ran a store in Northampton, would take in wheat, rye, Indian corn, pork, tallow and flax, butter, cheese, and tow cloth, but he would accept these items only for exchange. For others, Butler demanded cash or specified what items he would accept. In the town of Williamsburg, west of Northampton, Thomas Spafford offered six pence per pound for good flax, but he would also pay in "shoes, boots, indigo, and tobacco." *Hampshire Gazette*, Jan. 6, 1790.

50. Ibid., Nov. 10, 1790.

51. Ibid., May 3, June 5, 1793.

52. Ibid., Dec. 12, 1792.

53. *Federal Galaxy*, Brattleboro, Sept. 22, 1798.

54. *The Reporter*, Brattleboro, Aug. 1, 1803.

55. *Brattleborough Independent Freeholder and Republican Journal*, Dec. 8, 1808.

56. Ibid., March 6, 1808.

57. Ibid., Feb. 20, Dec. 8, 1808.

58. *Vermont Republican and American Yeoman*, Feb. 16, 1818.

59. Ibid., March 16, 1818.

60. Ibid.

61. Ibid., Feb. 16, 1818.

62. Merchant, *Ecological Revolutions*, 154–156. See also Conway Zirkle, "To Plow or Not to Plow: Comment on the Planters' Problem," *Agricultural History* 43, 1 (Jan. 1969): 87–89.

63. Merchant, *Ecological Revolutions*, 163, 189, 278, 279, 285; Clark, *Roots*, 146–158.

64. The *Connecticut Valley Farmer and Mechanic*, Springfield, May 1853, noted as late as that year that "the great farms are to be subdivided . . . [and farmers must] gain more from the subdivision than they ever gained from the whole."

65. See Merchant, *Ecological Revolutions*, 185–187, and Clark, *Roots*, 121–146, for a discussion of the demographic pressure on farm families.

66. Upset about this trend, one old farmer, Albert Comings, complained before the Connecticut River Valley Agricultural Society in 1853 that while earlier generations of farmers "paid out in money [for purchased goods] for his family the sum of $10.00, not less than $100.00 will now suffice." "Address before the Connecticut River Valley Agricultural Society, Lebanon, New Hampshire, Sept. 22, 1853," quoted in Wilson, *Hill Country*, 31.

67. In the eighteenth century and early nineteenth century, wheat flour had been a major export commodity from the valley, but nature conspired against wheat. In 1787, the Hessian Fly appeared in southern New England and began migrating north. Although newer wheat varieties reduced the problem of the Hessian fly, wheat crops continued to be plagued by "blast," or "rust." Dwight, *Travels*, 1:31. By the end of the eighteenth century, farmers

in the older towns in the lower valley lands had already shifted to rye. Judd, *Hadley*, 354, 355. Dwight noticed in the late eighteenth century that more and more rich alluvial land was being taken up raising "onions for commerce, particularly to the West Indies Islands." Dwight, *Travels*, 1:163. See also Lewis, "From Suffield to Saybrook, iii. Increasingly in the late eighteenth and nineteenth centuries lower valley farmers were also planting tobacco as a marketable export commodity.

68. Clark, *Roots*, 82–83.
69. *Hampshire Gazette*, Nov. 30, 1791.
70. Ibid., Oct. 30, 1793.
71. *The Reporter*, Brattleboro, Aug. 1, 1803.
72. *Vermont Republican and American Yeoman*, Feb. 16, 1818.
73. *Connecticut Valley Farmer and Mechanic*, Springfield, May 1853.

74. Dwight, *Travels*, 1:77. As late as 1853, the *Connecticut Valley Farmer and Mechanic* was complaining that particularly hill farmers still failed to practice scientific agriculture. The publication was dedicated to "bring[ing] science as largely as we can into companionship with agriculture." *Connecticut Valley Farmer and Mechanic*, Springfield, May 1853.

75. The *New England Farmer*, the mouthpiece of the improvementists, complained that upland farmers did not attend to modern scientific methods. *New England Farmer*, Mar. 15, 1851. I want to thank Richard Judd for this reference. Margaret Richard Pabst found that in the early years of the nineteenth century, hill farmers particularly had "little interest in the improvement of farming technique." Pabst, *Agricultural Trends*, 22, 19–23.

76. Judd Papers, "Miscellaneous, Vol. 14, 286, 287," 343; Mary Pepperrell Sparhawk Cutts, *Life and Times of Honorable William Jarvis of Weathersfield, Vermont by His Daughter* (New York, 1869), 337, 338, 339, 351. See Merchant, *Ecological Revolutions*, 203–217, for a discussion of the shift to improvement-oriented farming. For the role of gentlemen farmers in that shift, see Tamara Plakins Thornton, *Cultivating Gentlemen: The Meaning of Country Life among the Boston Elite, 1785–1860* (New Haven, 1989).

77. Lewis, "From Suffield to Saybrook," 210. See Diane Lindstrom, *Economic Development in the Philadelphia Region, 1810–1850* (New York, 1978), for a discussion of regional economies and economic development.

78. When Dwight traveled up the Connecticut River Valley in 1776, he found that "the most considerable manufacturing of duck and coarse linen cloth in the United States is established [in Northampton]." Dwight, *Travels*, 1:239.

79. *Hampshire Gazette*, Apr. 12, 1790. By 1892, one Boston factory was producing 2,000 yards a week. See William Weeden, *Economic and Social History of New England, 1620–1789* (Boston, 1890), 2:851.

80. Dwight, *Travels*, 1:31.

81. In 1811, William Jarvis introduced merino sheep into Vermont. Cutts, *William Jarvis*, 314, 315; Jerold Wikoff, *The Upper Valley: An Illustrated Tour along the Connecticut River before the Twentieth Century* (Chelsea, VT, 1985), 106.

82. Merchant, *Ecological Revolutions*, 191.
83. Dwight, *Travels*, 2:60.

84. Dwight, *Notes of a Northern Traveler*, 125. Major floods struck the region in 1801, 1811, 1831, 1850, 1854, and 1859. See *New England Farmer*, Sept. 1, 1848, for a discussion about the increase in flooding once the forest cover has been removed.

85. *Vermont Republican and American Yeoman*, Apr. 6, 1816; Lewis, "From Suffield to Saybrook," 107.

86. Judd, "Hadley, Vol. 3, 11," 351. See also Clark, *Roots*, 80–81.

87. Dwight, *Travels*, 1:77. Not only were animals more likely to be in enclosed pastures, but the number of animals and amount of pasture land grew dramatically from 1790 to 1830. Clark, *Roots*, 80, 81; Merchant, *Ecological Revolutions*, 278, 279, 280.

88. Dwight, *Travels*, 2:212. Judd notes that sowing ground cover began to be introduced into the central Connecticut River Valley in the late eighteenth century. Judd Papers, "Miscellaneous, 2, 162," 348, and "Miscellaneous, 13, 103," 352. See also Lewis, "From Suffield to Saybrook,"99. Dwight viewed the traditional farmers' farms as presenting a view of "uncouth and disgusting aspect of fields." Dwight, *Travels*, 2:83. With the development of crop rotation, deep plowing, and cleaned fields, Dwight felt that the "farms . . . are assuming a neater and more thrifty aspect." Dwight, *Travels*, 1:77.

89. Dwight, *Travels*, 1:255, 256.

90. Dwight, *Travels*, 2:212.

91. Patten, *Diary*. See also Judd Papers; Dwight, *Travels*; Lewis, "From Suffield to Saybrook," 64, 65; and Hayes, *History*, for the importance of mills for local farmers. "The history of the towns of the Connecticut Valley saw so many hardships in traveling long distances to mills, before they could be established nearer." Hayes, *History*, 112.

92. Hayes, *History*, 112; Charles Dean, "The Mills of Mill River," manuscript, Forbes Library, Northampton, MA, 1935; Judd, *Hadley*, 40.

93. *Seth Spring v. Sylvanus Lowell* (May 1805), MA Reports, Tyng, 1:422. Although the court ruled that Spring had a right to flood Lowell's fields, it also ruled that the jury should decide how long Spring could keep his dam up and at what height in terms of what "may be necessary." In 1831, in *Moses Fiske v. The Framingham Manufacturing Comp.* (Oct. 1831), MA Reports, Pickering, 12:68, Judge Shaw argued that "the public interest . . . coincides with that of the mill owner, and as the mill owner and the owner of the lands to be overflowed can not both enjoy their full rights without some interference, the latter shall yield to the former so that the former may keep up his mill and head of water."

94. *Inhabitants of Andover v. Ebenezer Sutton and Others* (Nov. 1846), MA Reports, Metcalf, 12:182.

95. *Benjamin French v. The Braintree Manufacturing Comp.* (Oct. 1839), MA Reports, Pickering, 23:216.

96. *Fiske v. Framingham*.

97. Wool was processed by fulling, which involved rough country woven wool being soaked and beaten by water-wheel-turned beaters to make the wool soft.

98. COPF, Sen. Doc. no. 8, *Report of the Commissioners, 1866* (Boston, 1866), 40.

99. JSC, "Report," 4. Hayes stated that fishing on the Connecticut "was one of the principal industries of the early settlers of Rockingham, Vermont." Hayes, *Connecticut River Valley*, 172.

100. Capt. Hooper of Walpole testified before the New Hampshire Fish Commission that before the dam at Turners Falls, shad and salmon ascended to the falls in such numbers that with a single haul of the seine 800 shad and 30 salmon were hauled in. NHCF, *Report Made to the Legislature of New Hampshire, 1870* (Concord, 1870), 2:4. Lyman Hayes claimed that as late as the 1840s, clerks and shopkeepers would close up, "posting signs on the door with the information, 'down at the eddy fishing.'" Hayes, *Connecticut River Valley*, 171.

101. COPF, *Report of the Commissioners, 1866*, 39.

102. In *Freary and Al. v. Cooke* (April 1779) reported in "Supplement" (1817), *MA Report*, Tyng, 14:488, the court ruled "there being in that case [a common fishery] no special property in the fish until they are caught," 490. See John Locke, *The Second Treatise on Civil Government*, chapter 5, paragraphs 25, 26, 29, 31, 36. See Keith Thomas, *Man and the Natural World: A History of the Modern Sensibility* (New York, 1983), 49.

103. Anadromous fish such as salmon live in different environments over their life span. Salmon spend most of their lives in the ocean, gaining ten to forty pounds in weight. Mature salmon then migrate upstream to the place they themselves were hatched. Once she has traveled to the spawning grounds, the typical female salmon will deposit eggs, which the male salmon will fertilize. Of the several thousand eggs (nineteenth-century commentators believed the Atlantic salmon deposited between 20,000 and 40,000 eggs), most either will fail to develop or will be eaten by hungry predators. About 10 percent of the eggs will hatch into healthy fry. These fry will spend the next few years gaining weight, until as young fingerlings, they will head out to sea. Only about 10 percent of the hatched fry will make it to the sea. Several years later and some ten to forty pounds heavier, half of those who made it out to sea will begin the trip back to the original spawning grounds. Shad follow a similar life pattern to salmon, though they do not need the same high DO level as salmon and will cast their eggs at their original spawning grounds above the tidal zone. Mature female shad produce three or four times more eggs than salmon.

104. R. A. Chapman, "Artificial Propagation of Fish," *Report* (Boston, 1857), 9. Arthur McEvoy, *The Fisherman's Problem: Ecology and Law in the California Fisheries, 1850–1980* (Cambridge, Eng., 1986).

105. MCIF, *Sixth Annual Report, 1872*, appendix, 1–5. The laws in the ancient charters specified: "It shall be free for any man to fish and fowl there [in a great pond, which was defined as any pond ten acres or more, even if the pond were wholly enclosed by a person's property], and may pass and repass on foot through any man's propriety for that end, so they trespass not upon any man's corn or meadow." Ancient Charters, p. 148, ch. 63. sec. 4. A 1641 law was amended in 1647 to give the right to fish and fowl to all free

men and specified that the right held not only in bays and navigable waters, but also in ponds over ten acres, and gave a person the right to cross another's property to get to the great pond. Judge Hoar noted in *Inhabitants of West Roxbury v. Stoddard* (Oct. 1863), MA Reports, Allen, 7:158, that the 1647 ordinance was "designed to establish a large and important public right." Hoar saw the purpose of these acts as "to declare a great principle of public right, to abolish the [English] forest laws, the game laws, and the laws designed to secure several and exclusive fisheries, and to make them all free." He read this act as making ponds "public property": "Ponds lie in common for public use."

106. Quoted from "Supplement" (Oct. 1857), MA Reports, Gray, 9:526. In *Freary and Al. v. Cooke*, 488, the court explained that under English, colonial, and American law there are different kinds of fisheries, one of which, common fisheries, is open to all but regulated by legislative act. There is no writ of trespass for a common fishery. Courts in this country and in England have ruled that "it is against common right, that fishing in public navigable rivers should belong to the crown or any individual" (490). *Black Commentaries* 2:39, 40. "By the charter of William and Mary, the property of the river is vested in the inhabitants of the province; and this the Court are bound to take notice of."

107. MCIF, *Sixth Annual Report, 1872*, appendix, 1, 2, 5.

108. Ibid.

109. Ibid., 6.

110. Ibid., 6, 7.

111. Ibid., 7–10. See also *Timothy Boutelle v. David Nourse* (May 1808), MA Reports, Tyng, 4:430; *Commonwealth v. John Wentworth and Others* (June 1818), MA Reports, Tyng, 15:187.

112. Chief Justice Shaw noted in reviewing the history of legislation on fisheries that "no one conversant with the legislation of Massachusetts, and who has witnessed the constant anxiety of the government for the salmon, shad, or alewives, and in regulating the fisheries of them, can doubt the [legislature's purpose in protecting fish]." *Commonwealth v. Essex Company* (June 1859), MA Reports, Gray, 13:239.

113. See MA Statute 1794, sec. 1. See also MA Statute 1791, sec. 1. In 1791, for example, the state passed a law "regulating the fishery in [the] Connecticut River" (sec. 1, 296). "No person or persons shall between the fifteenth day of March and the fifteenth day of June in any year, set or draw any seine or seines, or any other machine, for the purpose of catching fish in the Connecticut River, or in any river or stream falling into the same, from the rising of the sun on Saturday morning until the rising of the sun of Tuesday morning" (35). In 1794, the state outlawed fishing in or near the canal at South Hadley Falls, because the canal concentrated the fish migrations (48). This act was strengthened in 1813, further limiting when fishing for salmon and shad was allowed (107).

114. Ibid., 10–35. An 1804 act specified that "it is the duty of the owner or occupant of any such mill [or dam] to cause to be made and kept open a

sluice or passage-way for fish to pass up and down through the dam" (91, 133, 134).

115. *Commonwealth v. Knowlton.*

116. In *Commonwealth v. Knowlton*. In *Commonwealth v. Enoch Chapin* (Sept. 1827), MA Reports, Pickering 5:199, the court found Chapin's dam on the Connecticut River a violation of the fish acts because it restricted fish migration. And in 1829 in *Vinton v. Welsh*, the court ruled that the legislature had the right to force dams to be opened for fish migration. In the cases of limitations on fishing times and instruments, the courts upheld the right of the state and the towns by legislation from the state to restrict fishing. *Aaron Burnham v. Joseph Webster* (May 1809), MA Reports, Tyng, 5:267; *Commonwealth v. Wentworth.*

117. In Provincial Statute 8, Ann. c. 3, "all persons [we]re prohibited from placing in or across rivers or streams, any fixed implement or machine by which the free passage of fish may be obstructed." In Provincial Statute 15, Geo. 2, c. 6, "it [was] required of those who build dams across streams or rivers, to keep open, during a certain period, sluice-ways or passages for the fish to pass through." See *Commonwealth v. Chapin.*

118. See William Cowell's complaint in 1842 against Fisher Thayer, in which Cowell argues before the court that "according to custom of the country, a saw mill, or other mill has been kept up in the winter only." *William Cowell v. Fisher Thayer* (Nov. 1842), MA Reports, Metcalf, 5:253. See also *Timothy Hill and Wife v. Caleb Sayles* (Oct. 1849), Cushing, 4:549, in which Sayles was ordered by an earlier court decision to take up his dam gate so as not to flood fields between May 1 and November 1.

119. See Gary Kulik, "Dams, Fish, and Farmers: Defense of Public Rights in Eighteenth-Century Rhode Island," in Steven Hahn and Jonathan Prude, eds., *The Countryside in the Age of Capitalist Transformation: Essays in the Social History of Rural America* (Chapel Hill, 1987), 25–50; and Theodore Steinberg, *Nature Incorporated: Industrialization and the Waters of New England* (Cambridge, 1991).

120. Massachusetts, 1817, c151, *Vinton v. Welsh.*

121. *Staughton and Sharon, and Canton v. Edmund Barker and Daniel Vose* (Oct. 1808), MA Reports, Tyng, 4:524.

122. Although the legislature provided that dams that did not have sufficient fishways could be treated as nuisances, the court, with Justice Putnam dissenting, ruled that damages should be handled by suit for penalties, not by indictment for nuisance. *John Borden v. Samuel Crocker et al.* (Oct. 1830), MA Reports, Pickering, 10:383.

123. See George Rogers Taylor, *The Transportation Revolution, 1815–1869* (New York, 1951).

124. George Sheldon, "Old Time Traffic and Travel on the Connecticut," in *Proceedings of the Pocumtuck Valley Memorial Association, 1890–1898* (Deerfield, 1901), 3:122.

125. Sheldon, "Old Time Traffic." 117–129.

126. Hayes, *Connecticut River Valley*, 10.

127. *Hampshire Gazette,* Mar. 21, 1792.

128. Ibid., Jan. 2, 1793.

129. Judd Papers, "Hadley, Vol. 7, 170"; Dwight, *Travels,* 1:234.

130. Dwight, *Travels,* 2:243, 244; Hayes, *Connecticut River Valley,* 11.

131. Hayes, *Connecticut River Valley,* 12, 13. William Edmund Leuchtenburg, *Flood Control Politics: The Connecticut River Valley Problem, 1927–1950* (Cambridge, 1953), 16.

132. Sheldon, "Old Time Traffic," 129.

133. The courts ruled that under *Stowell v. Flagg* (a mill case), the act of incorporating the canal company allowed the canal company to compensate for damages those whose lands were affected by the building of the canal. Since the land owners were compensated, that process superseded ancient common law. *Jonathan Stevens v. The Proprietors of the Middlesex Canal Company* (Oct. 1815), MA Reports, Tyng, 12:465. See also *Proprietors of Sudbury Meadows v. The Proprietors of the Middlesex Canal* (Oct. 1839), MA Reports, Pickering, 23:36. Daniel Calhoun, *The American Civil Engineer: Origins and Conflicts* (Cambridge, MA, 1960).

134. Dwight, *Travels,* 1:235.

135. Ibid.

136. In 1796, petitions were brought to the general court concerning the dam obstructing the passage of shad. A proprietors committee was appointed to answer the petition. In 1797, John Strong appealed to the court of common pleas requesting fishing damages. In 1800, Jonathan Strong of Northampton sued the Proprietors for causing illness. He won in 1800, but on appeals, the supreme court overturned this ruling in 1803. Robert Bennett, "The Roots of the Holyoke Water Power Company" (Holyoke, 1985), manuscript, Holyoke Public Library, 1:48.

137. Judd Papers, "Miscellaneous, Vol. 11, 72," 408. In September of 1801 the Proprietors of the canal agreed to remove part of the dam. Bennett, "Roots of Holyoke," 1:49.

138. Judd, *Hadley,* 398; Dwight, *Travels,* 1:235, 236. Bennett, "Roots of Holyoke," 1:49, 50. See *Hampshire Gazette,* Jan. 6, 1791, and *Vermont Republican and American Yeoman,* Apr. 13, 1818, for an example of how these lotteries worked.

139. *Proprietors of Sudbury Meadows v. Proprietors of the Middlesex Canal.* In 1828, the canal company rebuilt its 1796 dam "higher, tighter, and broader, "raising water to cover more meadow longer. The company used the increased water level to sell power to mills. Although the new water level was higher than the original dam and the new level was used not for the canal at all but to power mills, the courts ruled against the meadow owners, arguing that allowing the canal company to use mill sites for power gave it the right to raise its dam.

140. Dwight, *Travels,* 1:1.

141. Ibid., 7. Dwight's optimism is reflected in his belief that cultivation will open up New England's beauty and improve its weather (39).

142. Dwight, *Travels,* 4:117.

143. Ibid., 4:38. See also *The Palmer Co. Petitioners v. Isaac Ferrill* (Sept. 1835), MA Reports, Pickering, 17:58.

CHAPTER 2

1. Although developmental economists have argued that traditionalist farmers were more destructive to the environment than were market-oriented farmers who used more modern agricultural techniques, it is a mistake to overcredit market-oriented farmers as being more environmentally sensitive because of their long-term economic interests. Individually, traditionalists as well as market-oriented farmers may have acted in ways that were destructive to the environment. See Carolyn Merchant, *Ecological Revolutions: Nature, Gender, and Science in New England* (Chapel Hill, 1989), 154–156. See also Donald Worster, *The Dust Bowl: The Southern Plains in the 1930s* (Oxford, 1979).

2. See Christopher Clark, *The Roots of Rural Capitalism: Western Massachusetts, 1780–1860* (Ithaca, 1990). See also Timothy Dwight, *Travels Through New England and New York* (Cambridge, 1969), 4:348.

3. Although Levi Shepard represented a new focus in the valley, employing labor to manufacture goods for distant markets, his technology was very familiar. Shepard built his factory at the back of his home lot. All the machinery was worked by hand, with a boy or girl hired to turn the wheel, yet Shepard's market was not local.

4. *Hampshire Gazette*, Sept. 29, 1790.

5. See Jonathan Prude, *The Coming of Industrial Order: A Study of Town and Factory Life In Rural Massachusetts, 1813–1860* (Cambridge, 1983), for an example of the impact of mills on the surrounding countryside.

6. Charles J. Dean, "The Mills of Mill River," manuscript, 1935, Forbes Library, Northampton, MA, 21.

7. Besides carding, fulling, sawing, and grinding, some local millers were also manufacturing tools for local farmers. Rufus Hyde and his sons built a mill dam to power a trip-hammer. The Hydes used the trip-hammer to hammer out axes and scythes. They also used the waterwheel to turn their grindstones that sharpened axes sold to local farmers. Ibid., 23.

8. Vera Shlakman, *Economic History of a Factory Town: A Study of Chicopee, Massachusetts, Smith College Studies in History*, vol. 20, nos. 1–4 (Northampton, 1934–1935), 22.

9. Dean, "Mills of Mill River," 30–33.

10. *Stephan Cook v. William Hull* (Oct. 1825), MA Reports, Pickering, 3:269; *Isaac Biglow and Others v. Mellen Battle and Others* (Oct. 1818), MA Reports, Tyng,15:313.

11. *Biglow v. Battle*; Dean, "Mills of Mill River," 30, 31.

12. Dwight, *Travels*, 2:197, 4:342, and 4:343–346.

13. Theodore Dwight, *Notes of a Northern Traveler* (Hartford, 1831), 97, 831.

14. Ibid., 94; Dwight, *Travels*, 2:195; Alexander Johnson, *Connecticut: A Study of a Commonwealth-Democracy* (Boston, 1895), 357, 358, 363.

15. Edwin M. Bacon, *The Connecticut River and the Valley of the Connecticut: Three Hundred and Fifty Miles from Mountain to Sea, Historical and Descriptive* (New York, 1906), 430.

16. Edmund Dwight's family had been merchants and investors in the early canal projects along the Connecticut. Edmund, married to an Eliot, moved to Boston in 1816 and joined the inner circle of wealthy Bostonians just expanding into investments in textile production. See Shlakman, *Economic History*, 27–47.

17. By 1841, Chicopee had the Chicopee, Cabot, Perkins, and Dwight manufacturing corporations, employing 2,500 hands. The town also had a number of skilled workers in the Belcher Iron Works and the Ames Manufacturing Company, which produced metal tools, cutlery, knives, and swords. Shlakman, *Economic History*, 25–36; see also Bacon, *Connecticut River*, 423.

18. Shlakman, *Economic History*, 27.

19. Will L. Clark, ed., *Western Massachusetts: A History, 1636–1925* (New York, 1926), 2:946, 947, 971.

20. Lyman S. Hayes, *The Connecticut River Valley in Southern Vermont and New Hampshire: Historical Sketches* (Rutland, 1929), 290, 292, 293. See R. H. Clapperton, *The Paper Making Machine: Its Invention, Evolution, and Development* (Oxford, 1967).

21. Robert Bennett, "The Roots of the Holyoke Water Power Company" (Holyoke, 1985) manuscript, Holyoke Public Library, 2:7.

22. Silk manufacturing continued in several towns along the central and lower valleys. Bacon, *Connecticut River*, 418. Dean, "Mills of Mill River," 54–58.

23. Harold Wilson, *The Hill Country of Northern New England: Its Social and Economic History* (New York, 1967), 145.

24. On the Franconia ironworks, see Jerold Wikoff, *The Upper Valley: An Illustrated Tour along the Connecticut River before the Twentieth Century* (Chelsea, VT, 1985), 118, 119, 120; Dwight, *Travels*, 4:117. On the locating of ironworks on water-power sites, see *Paul Sibley and another v. Thomas Hoar and another* (Oct. 1855) MA Reports, Gray 4:222.

25. There were almost a million and a half sheep in Vermont and over a half million in New Hampshire by the 1830s.

26. Wikoff, *Upper Valley*, 123.

27. Margaret Richards Pabst, "Agricultural Trends in the Connecticut Valley Region of Massachusetts, 1800–1900," *Smith College Studies in History*, vol. 26, nos. 1–4 (Northampton, Oct. 1940–July 1941), 53, 54. A history of Ryegate, Vermont, at the turn of the century noted that "before 1850 woolen mills could be found in [many of northern Vermont's towns]." Quoted in Wilson, *Hill Country*, 45.

28. Dean, "Mills of Mill River," 33.

29. Valley farmers continued to raise wheat and grains, but their relative importance to the region's agriculture declined after 1840, with production itself declining after 1850. Dairy products jumped in value over the first half of the nineteenth century. Pabst, *Agricalatural Trends*, 54. This shift to perish-

able vegetables, fruits, and dairy products was encouraged after the opening up of the Erie Canal and the penetration into New England and New England's traditional markets of staple agricultural goods—particularly grain—from central New York and the upper Midwest.

30. The impact on gender roles of the shift of vegetables and dairy products from home use to market produce is an area of research that needs more work. Carolyn Merchant's *Ecological Revolutions* explores some of these issues.

31. Quoted in Pabst, *Agricultural Trends*, 53.

32. In 1885, when Massachusetts took a census of its agricultural products, it found that for farm communities around the industrial centers, dairy products followed by garden vegetables accounted for the majority of farm income. Hampton County had a total of $175,571 in meats and game, $234,264 in cereals, $302,426 in vegetables, $950,208 in dairy, and $126,021 in poultry. Hampshire County had a total of $210,689 in meats and game, $228,341 in cereals, $266,978 in vegetables, $1,050, 825 in dairy (mostly milk), and $111,843 in poultry. The produce of individual towns reflected a dramatic shift toward dairy, mostly milk, and toward vegetables, particularly cabbage, potatoes, and squashes, and to a lesser extent toward poultry. State of Massachusetts Bureau of Labor Statistics, *Census of Massachusetts, 1885, vol. 3, Agricultural Products and Property* (Boston, 1887), 568, 569, 570, 571, 216–294 (hereafter, *MBLS, Census*).

33. Hadley Falls Company, "Report of the History and Present Condition of the Hadley Falls Company at Holyoke, MA," 1853, manuscript, Holyoke Public Library, 19 (hereafter, "Hadley Falls Company Report").

34. Quoted in Pabst, *Agricultural Trends*, 26.

35. Despite the Palmer Company's argument that the mill would bring benefits for Ferrill, the court found against Palmer. The court ruled that although it was true that the dam did bring economic growth, the benefits of "the general prosperity of the settlement are too contingent, remote, and indirect." Even more significant for the future, the court, having more foresight than the Palmer Company, argued that just as no land owner can claim damages to the value of land because of "injury arising from manufacturers which would occasion noxious smells, or uncomfortable voices, or other means of annoyance to a neighborhood by attracting a bad population, increased taxation, or causing pauperism and mendacity," the manufacturers couldn't claim the privilege of the potential prosperity their factory might bring. *The Palmer Co. v. Isaac Ferrill* (Sept. 1835), MA Reports, Pickering, 17:58.

36. The hill town of Pelham, for example, was producing $17,558 worth of wood products in 1885 and only $17,278 in dairy and $5,438 in vegetables. Northampton was producing only $9,974 in wood products but was producing $74,679 in dairy products, and $27,020 in vegetable produce. Upland farms continued to produce orchard fruits even though these did not compete with the produce from lowland farms. Pelham produced $4,019 in fruits, more than it was producing in grains at $2,699 but significantly less than Northampton's $8,479. MBLS, *Census*, 570–571. Pelham also experienced a significant loss

in population over the nineteenth century. In 1850, Pelham had 983 residents, in 1860, 748, in 1870, 673, and by the end of the century, only 462. Bureau of the Census (Wash. D.C.) *U.S. Census of Population*, 1850, 1860, 1870, 1900. See also Pabst, *Agricultural Trends*, 16, 17, 40, 47.

37. By midcentury, New Hampshire was sending almost 20 million feet of timber down the river. Wikoff, *Upper Valley*, 64; Dwight, *Travels*, 2:72; Roland Harper, "Changes in the Forest Area of New England In Three Centuries," *Journal of Forestry* 16 (Apr. 1918): 443, 449. At Hartland, Vermont, David Summer erected extensive lumber mills and sent cut lumber downriver to Massachusetts and Connecticut towns, where he had lumber yards to store and sell his product. Hayes, *Connecticut River Valley*, 41.

38. "Hadley Falls Company Report" 20.

39. Harper, "Changes in the Forest Area," 447. Increasingly after 1840, railroads began consuming massive amounts of the region's lumber both for construction and for fuel. In 1846, G. B. Emerson in his classic report to the State of Massachusetts on its trees and shrubs noted that the 560 miles of railroad in the state were consuming 53,710 cords of wood annually. "Hadley Falls Company Report," 449.

40. Edward Everett Hale, *Tarry at Home Travels*, (New York, 1908) 111.

41. Quoted in Harper, "Changes in the Forest Area," 449.

42. By 1880, the New Hampshire Board of Agriculture noted that railroads at Concord alone consumed "70,000 cords of lumber and 1,000 of ties." Quoted in Wilson, *Hill Country*, 6.

43. "Hadley Falls Company Report," 20; Wilson, *Hill Country*, 47.

44. Vermont's forests dropped from 60 to 45 percent of the state's land area between 1820 and 1850 and to 35 percent by 1880. Harper, "Changes in the Forest Area," 449; See *Henry Irvine v. Daniel Stone* (Nov. 1850) MA Reports, Cushing, 6:508, for a case involving bringing coal to Boston to be used as a heating fuel. The cost of shipping coal from Philadelphia, where it was brought in from the mines northwest of the city, to Boston was $3.75 a ton for stove coal plus $735 freight for 202 tons of coal. At these prices, coal soon became competitive with northern cord wood. By midcentury, most New England cities had coal yards that sold to local businesses. *Commonwealth v George E. Mann* (Oct. 1855) MA Reports, Grey, 4:212.

45. George Perkins Marsh, "Report, Made under Authority of the Legislature of Vermont on the Artificial Propagation of Fish" (Montpelier, 1857) 14.

46. MSBH, *Fourth Annual Report of the State Board of Health, 1872* (Boston, 1873), 107.

47. Wikoff, *Upper Valley*, 65. For a description of this process in the northern Great Lakes, see William Cronon, *Nature's Metropolis: Chicago and the Great West* (New York, 1991).

48. Robert Bennett, "History of the Holyoke Water Power Company," manuscript (Holyoke Public Lib, 1989) 262, 277.

49. After the introduction of the railroad into the northern valley, dairy specialization increased, and the "milkshed" for Boston moved northward. Between 1850 and 1880, the number of dairy cows in Vermont rose from

146,128 to 217,033. See Wilson, *Hill Country,* 301–326, for a discussion of the development of specialized dairy farming in northern New England.

50. Wikoff, *Upper Valley,* 106; Wilson, *Hill Country,* 45, 145, which also reports that as a consequence of the decline of these small manufacturing shops, factory villages began to lose populations. Sheffield, for example, which in 1830 had a variety of manufacturing enterprises, by the end of the century had only a sleepy gristmill and was losing population. See Lewis Stilwell, *Migration from Vermont* (Montpelier, 1948), for a discussion of the loss of population in small manufacturing villages and hill country farms.

51. *Boston Herald,* July 22, 1882; Lyman Scrapbooks.

52. "Hadley Falls Company Report," 5.

53. Robert F. Dalzell, Jr, *Enterprising Elite: The Boston Associates and the World They Made* (Cambridge, 1987).

54. "Hadley Falls Company Report," 5.

55. The new company was incorporated by the state to construct a dam across the Connecticut River, and locks and canals, and "by creating water-power, to be used by said corporation for manufacturing articles from cotton, wool, iron, wood, and other material, and to be sold or leased to other persons and corporations to be used for manufacturing or mechanical purposes." Ibid. 5. See also Bennett, "Roots of Holyoke," 2:11; Constance McLaughlin Green, *Holyoke, Massachusetts: A Case History of the Industrial Revolution in America* (New Haven, 1939) 20. Among the investors in the new company were Alfred Smith of the old Hadley Falls Company; Samuel Cabot; William Appleton; Erastus Bigelow, the carpet loom manufacturer; Theodore Lyman, who told his son that without manufacturing New England would starve; George Lyman; and the local Dwight family, who had investments in most of the region's manufacturing enterprises. Bennett, "Roots of Holyoke," 1:65.

56. On Ely, see "Picturesque Hamden," quoted in Bennett, "Roots of Holyoke." Once the company was fully incorporated, its capital stock was fixed at $4 million. "Hadley Falls Company Report," 5.

57. The company calculated "a mill power—estimated at 60 or 70 horse power—[a]s 30 cubic feet of water per second over 25 foot fall." Ibid., 16. See also Bennett, "History" 104, for a discussion of mill power.

58. "Hadley Falls Company Report," 3, 4, 16. At Holyoke, mill power was defined as enough water to power 3,584 spindles for cotton cloth for sixteen hours a day. Green, *Holyoke, Massachusetts,* 21. For a discussion of the concept of mill power, see Theodore Steinberg, *Nature Incorporated: Industrialization and Waters of New England* (Cambridge, 1991), 83–86.

59. "Hadley Falls Company Report ," 5, 10, 11, 13, 22, 29, 30, 76.

60. Quoted in Bacon, *Connecticut River,* 421. See also Green, *Holyoke, Massachusetts,* 28. The loss of the dam at Hadley Falls was costly but not particularly tragic. That was not the case for the dam in Williamsburg just to the north, where the Mill River and Williamsburg Reservoir Company" built in 1866 a huge dam and reservoir system on the Mill River. On May 16, 1874, after a series of heavy rains, the dam gave way. A towering wall of water spilled downriver, wiping out mills, homes, and lives. Dean, "Mills of Mill River."

61. Holyoke Water Power Company, "Connecticut River Flood at Holyoke, 1927," manuscript, Holyoke Public Library; "Hadley Falls Company Report," 7, 8. The new dam was begun in 1900.

62. "Hadley Falls Company Report," 11.

63. Ibid.

64. Green, *Holyoke, Massachusetts*, 30.

65. Ibid., 62–65.

66. Bennett, "*History*," 78.

67. Charles Bryan, *Paper Making as Conducted in Western Massachusetts* (Springfield, 1874), 59.

68. Ibid. South Hadley Falls had another two companies that employed 225 workers, while West Springfield's three companies employed another 235 workers.

69. To make quality paper, rags had to be sorted, cut, bleached, washed, then boiled. The fiber was then put into water-powered beaters with clay and rosin added as adhesives to make the stuff. The stuff was fed into Jordon engines for further processing and then pumped into the Fourdrinier machines, where it was formed into paper by being run through rollers and driers to extract the last of the water.

70. Clark, *Western Massachusetts*, vol. 2, 909–916; Green, *Holyoke, Massachusetts*, 72, 93.

71. Bennett, "History" 210, 103, 104. Clemens Herschel developed the measuring device, or testing flume, in 1881, which, when attached to the waterwheels measured the flow of water through the wheel. The new charging system concerned paper manufacturers, but even with the new charging system, the water power at Holyoke was four times cheaper than comparative steam power costs.

72. Lyman Diaries, June 28, 1867.

73. Bennett, "Roots of Holyoke," 1:72–74. For a discussion of the development of the industry at Holyoke, see Green, *Holyoke, Massachusetts*.

74. "Hadley Falls Company Report," 22.

75. See John T. Cumbler, "Whatever Happened to Industrial Waste: Reform, Compromise, and Science in Southern New England," *Journal of Social History* (Winter 1995), for a discussion of waste in nineteenth-century New England.

76. MSBH, *Third Annual Report of the State Board of Health, 1871* (Boston, 1872), 60–61.

77. Massachusetts Bureau of Labor Statistics, *Sixth Annual Report, 1875* (Boston, 1875), 392.

78. "Hadley Falls Company Report,"12.

79. Ibid., 13.

80. Quoted in Green, *Holyoke, Massachusetts*, 42.

81. Ibid.

82. Massachusetts Bureau of Labor Statistics, *Sixth Annual Report, 1875*, 392.

83. Clark, *Western Massachusetts*, 2:918, 919.

84. The court case was Delotes Lumbard v. Charles Stearns (Sept. 1849), MA Reports, Cushing 4:60. On the Holyoke system, see Green, *Holyoke, Massachusetts*, 120.

85. In the 1840s, Chicopee, whose residents got their drinking water from wells beside their homes, suffered an epidemic of typhoid. The typhoid outbreak finally led the city to build a water system, which brought water into the city "from the hill above the town." *Springfield Republican*, May 8, 1901, quoted in Shlakman, *Economic History*, 54.

86. MSBH, *Fourth Annual Report*, 32.

CHAPTER 3

1. James Olcott, speech before the Agricultural Board of Connecticut, reprinted in CSBH, *Ninth Annual Report of the State Board of Health* (Hartford, 1887), 239, 241, 242.

2. Ibid., 239.

3. MSBH, *Eighth Annual Report of the State Board of Health* (Boston, 1876), 403.

4. NHSBH, *Sixth Annual Report* (Concord, 1887), 6:188.

5. Donald Worster, *Nature's Economy: A History of Ecological Ideas* (Cambridge, 1985).

6. Such an analysis called for a more visible hand, specifically that of the state.

7. Quoted in Vera Shlakman, *Economic History of a Factory Town: A Study of Chicopee, Massachusetts*, Smith College Studies in History, vol. 20, nos.1–4 (Northampton, 1934–1935), 22.

8. NHSBH, *Third Annual Report* (Concord, 1884), 3:279. Crosby figured that sixteen gallons of water would be used per person daily, with nine to wash the sewage away. NHSBH, *First Annual Report* (Concord, 1882), 1:242–243.

9. CSBH, *Fifth Annual Report of the State Board of Health* (Hartford, 1883), 132.

10. Ibid., 134.

11. *City of Boston v. Jesse Shaw* (Mar. 1840), MA Reports, Metcalf, 1:130. *Samuel Downer and Others v. City of Boston* (Mar. 1851), MA Reports, Cushing, 7:277.

12. In 1876, the Massachusetts State Board of Health divided the sewer systems of towns into those that drained into salt water bays; those that dumped into tidal streams; and those that drained into freshwater streams, a majority of which drained into the Connecticut. MSBH, *Eighth Annual Report*, 203. In 1844, Hartford built a system that emptied into either the Park River or the Connecticut. CSBH, *Second Annual Report of the State Board of Health* (Hartford, 1880), 72. Because most of the larger coastal cities dumped sewage into tidal streams and utilized relatively unpolluted upland sources for their water supplies, in 1870 when the newer industrial cities were just beginning to establish themselves, the Massachusetts State Board of Health felt

that typhoid was a disease of small towns, not of larger cities. That evaluation would change within a few short years. MSBH, *Second Annual Report of the State Board of Health, 1870* (Boston, 1871), 118. See Fern Nessen, *Great Waters: A History of Boston's Water Supply* (Hanover, 1983), for a discussion of Boston's search for clean water.

13. CSBH, *Fifth Annual Report*, 135.

14. CSBH, *Tenth Annual Report of the State Board of Health* (Hartford, 1888), 203.

15. MSBH, *Eighth Annual Report*, 207.

16. The term *water-closet* was used in the late nineteenth century interchangeably with *privy*. Sometimes it denoted a flush toilet, but not always. MSBH, *Eighth Annual Report*, 109, 118.

17. The city of Lowell built sewers in 1842 and extended them in 1871. Until the 1860s, most of the runoff was rainwater, but increasingly after 1871, it was sewage. In 1884, the Middlesex Corporation successfully sued the city to stop the sewers from discharging into its millpond. The city then switched to dumping directly into the Merrimack and Concord Rivers.

18. MSBH, *Fourth Annual Report of the State Board of Health, 1872* (Boston, 1873), 32.

19. *William Wright v. the City of Boston* (Mar. 1852), MA Reports, Cushing, 9:233.

20. Hadley Falls Company, "Report of the History and Present Condition of the Hadley Falls Company, at Holyoke, MA," 1853, report to the stockholders, manuscript, Holyoke Public Library, Holyoke, MA, 19.

21. MSBH, *Eighth Annual Report*, 207; Constance McLaughlin Green, *Holyoke, Massachusetts: A Case History of the Industrial Revolution in America* (New Haven, 1939), 120.

22. As the Connecticut State Board of Health noted, "Like most of the older cities . . . sources of water, air, and soil pollutions detrimental to health have been caused as a busy, thriving manufacturing city was developed from the straggling irregular collection of buildings that first marked the site." CSBH, *Second Annual Report*, 72.

23. MSBH, *Fourth Annual Report*, 40. See also NHSBH, *Third Annual Report* (Concord, 1884), 3:296.

24. NHSBH, *Fourth Annual Report*, 4:224.

25. CSBH, *Tenth Annual Report*, 260.

26. CSBH, *Second Annual Report*, 20, 137, 139.

27. Ibid. CSBH, *Ninth Annual Report*, 20.

28. CSBH, *Tenth Annual Report*, 260.

29. CSBH, *Ninth Annual Report*, 239, 241, 242.

30. See Theodore Steinberg, *Nature Incorporated: The Industrialization of the Waters of New England* (Cambridge, 1991), for discussions of the pollution of the Charles and the Merrimack Rivers.

31. MSBH, *Fourth Annual Report*, 81.

32. MSBH, *Eighth Annual Report*, 53. For each 100 pounds of raw material, the wood pulp process generated 63 pounds of liquid organic refuse

and 37 pounds of useful fiber; 100 pounds of white rags produced 6 pounds of liquid organic refuse and 85 pounds of fiber (51).

33. MSBH, *Eighth Annual Report*, 53.
34. CSBH, *Tenth Annual Report*, 192. MSBH, *Eighth Annual Report*, 56, 57.
35. Ibid., 51.
36. Ibid., 53.
37. Green, *Holyoke, Massachusetts*, 147, 155.
38. Walter Hurd, *The Connecticut* (New York, 1947), 40; Edwin M. Bacon, *The Connecticut River and the Valley of the Connecticut: Three Hundred and Fifty Miles from Mountain to Sea, Historical and Descriptive* (New York, 1906), 354.
39. Edward Everett Hale, *Tarry at Home Travels* (New York, 1906), 93. Although the actual acreage of New England forests did not continue to decline after the 1880s, what Hale perceived was the impact of a new, more aggressive clear-cutting of the region's conifers.
40. CSBH, *Tenth Annual Report*, 192–194.
41. MSBH, *Ninth Annual Report, of the State Board of Health, 1878* (Boston, 1879), 60.
42. Ibid., 195.
43. Four hundred pounds of washed wool would take sixty pounds of dyewood, fifty pounds of alum, and ten pounds of potash.
44. MSBH, *Ninth Annual Report*, 196.
45. Processing 30 tons (67,200 pounds) of wool generates wastes of 8–10 tons of grease and dirt, 14–15 tons of urine, 2 tons of oil, 2 tons of pigs' dung, 2 tons of pigs' blood, 25 tons of urine for the second wash, one ton of soda, over 2 tons of soap, 2 tons of fuller's earth, 20 tons of dye stuff, and 2 tons of alum. MSBH, *Eighth Annual Report*, 39, 40, 41.
46. Ibid., 44.
47. Ibid., 42.
48. Ibid.
49. Ibid., 43.
50. CSBH, *Tenth Annual Report*, 199.
51. NHSBH, *First Annual Report* (Concord,1882), 1:54, 56.
52. MSBH, *Eighth Annual Report*, 63.
53. Ibid., 188.
54. CSBH, *Tenth Annual Report*, 203.
55. Ibid., 191.
56. CSBH, *Thirteenth Annual Report of the State Board of Health* (Hartford, 1891), 219.
57. Ibid., 221.
58. CSBH, *Tenth Annual Report*, 21.
59. NHSBH, *First Annual Report*, 1:57.
60. Lyman Diaries, July 10, 1865.
61. MSBH, *Eighth Annual Report*, 20.
62. CSBH, *First Annual Report of the State Board of Health* (Hartford, 1879), 61. In 1869, the Massachusetts State Board of Health asserted that "the agency of foul and putrid air . . . in causing disease is a very recent dis-

covery, yet nothing is better established." *Eighth Annual Report*, 51. New Hampshire and Connecticut were reluctant to give up the miasma theory even after the works of Lister, Koch, Henle, and Pasteur had shown the role of germs. "Foul water . . . generates miasma of disease and . . . poisonous vapors." NHSBH, *First Annual Report*, 1:75; CSBH, *Twelfth Annual Report of the State Board of Health* (Hartford, 1890), 29.

63. NHSBH, *First Annual Report*, 1:8; CSBH, *Tenth Annual Report*, 40, 41, 179, and *Fourth Annual Report of the State Board of Health* (Hartford, 1882), 28.

64. Charles Chapin, *A Report on State Public Health Work* (Chicago, 1916), 61. By the 1840s, more and more doctors were noticing the connection between poor sanitary conditions and poor health. Differing mortality rates between rural and urban populations and the simple commonsense connection between the smell associated with filthy environments and high disease rates led to the belief that the environment itself caused disease.

65. Ibid., 21.

66. CSBH, *Tenth Annual Report*, 312.

67. MSBH, *Eighth Annual Report*, 63.

68. MSBH, *Ninth Annual Report*, 59.

69. CSBH, *Fourteenth Annual Report*, 436. In 1890, the city of Hartford alone was using five million gallons of water daily, which was then flushed out into the Connecticut.

70. CSHB, *Ninth Annual Report*, 239.

71. Ibid., 241.

72. Ibid., 240

73. See Robert Gottlieb, *Forcing the Spring: The Transformation of the American Environmental Movement* (Washington, DC, 1993), for a discussion of how urban health reformers pioneered in their concern for the environment.

CHAPTER 4

1. Timothy Dwight, *Travels Through New England and New York* (Cambridge, 1969), 4:117; see also 4: 338–350; Theodore Dwight, *Notes of a Northern Traveler* (Hartford, 1831), 97, 111.

2. In 1794, when the commercial leaders around Springfield, Hadley, and Northampton began building a new dam and canal around the falls on the Connecticut River, local fishers protested to "the gentlemen and proprietors of the locks and canals." But the protest did not stop the building of the dam. *Hampshire Gazette*, Jan. 15, 1794. In 1796, the fishers appealed to the legislature for relief, and in 1797, John Strong appealed to the court of common pleas requesting damages for loss of fishing. Robert Bennett, "The Roots of the Holyoke Water Power Co.," manuscript, 1985, Holyoke Public Library, Holyoke, MA, 1:48. In 1819, Ariel Cooley (Enoch Chapin's uncle) avoided a suit for blocking fish migration by building a fishway around his dam at Hadley Falls.

3. Alternatively, these angry fishers and farmers might take action in the court of common pleas for damages against the dam owners based on their common-law rights of property.

4. David Daggett, "A Brief Account of a Trial at Law, in which the Influence of Water Raised by a Mill-dam, on the Health of the Inhabitants in the Neighborhood Was Considered," *Memoirs of the Connecticut Academy of Arts and Science* (New Haven, 1813), vol. 1, pt. 1, no. 12, 131–134.

5. *Horace White v. Oliver Moseley et. al* (Sept. 1827), MA Reports, Pickering, vol. 5.

6. Thoreau identified with the fish that could no longer make their migration upstream. "I for one am with thee." Henry David Thoreau, *A Week on the Concord and Merrimack Rivers* (New York, 1966), 40.

7. The Mill Acts followed earlier legislation by the colonial general court.

8. In 1824, the legislature also amended the act to include land down from the mill (Statute 1824 c153, 3).

9. See Morton Horwitz, *The Transformation of American Law, 1780–1860* (Cambridge, 1977), 2–12, 21–22, for a discussion of U.S. common law. See also William Nelson, *Americanization of the Common Law: The Impact of Legal Change on Massachusetts Society, 1760–1830* (Athens, 1994), xi, 8–10. As Horwitz and Nelson have shown, common law not only became Americanized, it also became an instrument in the hands of activist judges for making law. Horwitz, *Transformation of American Law*, 27, 30; Nelson, *Americanization of the Common Law*, 169–172. Horwitz argues that increasingly, judges saw their decisions in terms of the public good. Common law for these judges was not meant to hinder economic development. See also Leonard Levy, *The Law of the Commonwealth and Chief Justice Shaw* (Cambridge, 1957).

10. *Seth Spring v. Sylvanus Lowell* (May, 1805), MA Reports, Williams, 1:422.

11. *Abel Stowell v. Samuel Flagg* (Sept. 1814), MA Reports, Tyng, 9:364. Parker argued that the 1796 act that is referred to as Statute 1795 c74 removed the actions of mill owners from common law. In 1815, the court ruled that when the legislature incorporated the canal company, it allowed the company to compensate land owners for damages; thus the legislature superseded action at common law. *Jonathan Stevens v. The Proprietors of the Middlesex Canal Company* (Oct. 1815), MA Reports, Tyng, 12:465.

12. *Commonwealth v. Jonathan Knowlton* (June 1807), MA Reports, Tyng, 2:529.

13. *Stoughton and Sharon, and Canton v. Edmund Barker and Daniel Vase* (Oct. 1808), MA Reports, Tyng, 4:524.

14. Ibid. See Nelson, *Americanization of the Common Law*, especially 159–160; Horwitz, *Transformation of American Law*; Charles Harr and Barbara Gordon, "Riparian Water Rights vs. A Prior Appropriative System: A Comparison," *Boston Law Review* 1957, 207–255, and Steinberg, *Nature Incorporated*.

15. *Inhabitants of Watertown v. Daniel Draper* (Oct. 1826), MA Reports, Pickering, 4:165. Henry David Thoreau asked in 1839 who was there to represent the fish before the legislature, or, as he put it, "who hears the fish when they cry?" *A Week*, 40.

16. Bennett, "Roots of Holyoke," 1:51.

17. The court did find that Chapin's dam was in violation of the fish acts. Justice Parker believed that the remedy for the obstruction of fish migrations was in statutory laws and enforcement. *Commonwealth v. Chapin*, 5:199.

18. Bennett, "Roots of Holyoke," 1:51–53.

19. In *Commonwealth v. Knowlton*, 529, the courts ruled that Knowlton's dam violated statutory law. In *Borden v. Crocker*, the court noted that damages should be handled by suit for penalties, not by indictment for a nuisance under common law. *John Borden v. Samuel Crocker et al.* (Oct. 1830), MA Reports, Pickering, 10:383.

20. *Bavil Seymour v. William Carter* (Sept. 1841), MA Reports, Metcalf, 2:520.

21. Thoreau, *A Week*, 35.

22. Nelson, *Americanization of the Common Law*, 53.

23. *Wolcott Woollen Manufacturing Comp. et al., v. Jacob Upham* (Sept. 1827), MA Reports, Pickering, 5:292. Yet all was not completely smooth sailing for the manufacturing interests of the state. In 1830, shortly after Parker's untimely death, Jeduthum Stevens and his partners, who had erected a mill dam in Brookfield in 1826 to power their textile mill, found themselves in court. Stevens argued before the court that "it had been the policy of this commonwealth to bring into action as much mill power as the Capital and business of its citizens require." But unfortunately for Stevens, Justice Putnam in dissenting opinions had reflected an older vision that was more suspicious of the argument of corporations; Putnam ruled that the public good intended was for the benefit of local community, not for the interests of capital and business. *Commonwealth v. Jeduthum Stevens* (Oct. 1830), Pickering, 10:247. Reflecting the more traditional economic world, the court ruled that the 1795 c74 act was to "encourage the building of mills serviceable for the public good and the benefit of the town and probably no conflict was anticipated between public spirited proprietors of mills having these objects in view and those who might represent and manage other public interests."

24. See Levy, *The Law*.

25. *Silas Bemis v. Samuel Clark* (Oct. 1831), MA Reports, Pickering, 12:452. In that same year, Shaw argued in *Moses Fiske v. The Framingham Manufacturing Comp.* (Oct. 1831), MA Reports, Pickering, 12:68, that "the public interest in such case coincides with that of the mill owner." See also *Benjamin French v. The Braintree Manufacturing Comp.* (Oct. 1839), MA Reports, Pickering, 13:216.

26. *William Ashley v. Harlow Pease* (Sept. 1836), MA Reports, Pickering, 17:268. This case also reflects how shifts in technology and market forces can affect the land. Pease originally ran a fulling mill for the local farm families to bring in their woven wool to be fulled. The fulling mill only ran for a part of the year, as Ashley noted in court: "The business of fulling cloth at the fulling mill had never required the use of the water for a greater period than 20 weeks yearly, but that the defendant had used the water for the carding machine during nearly the whole of the year." As a result, Ashley's meadows were never able to dry out enough to grow hay. When in 1845 the Lancashire Corporation built a dam on the Nashua River that cut off the water supply for periods of June and July to James Pitts's small gristmill, Pitts went to court to argue that he had prior rights as an ancient holder of a mill site privilege. Lancaster argued that it was using the water "reasonably," which qualified

Pitts's use. Judge Shaw argued that "what is a reasonable use must depend on circumstances; such as . . . the state of improvement in manufacturing and the useful arts." Shaw ruled for the Lancaster Mills. *James Pitts v. the Lancaster Mills* (Oct. 1847), MA Reports, Metcalf, vol. 13. 320 Shaw ruled in *Cary v. Daniels* that a mill owner held the right of reasonable use of the water. "One of the beneficial uses of a water course and in this country one of the most important is its application to the working of mills and machinery, a use profitable to the owner and beneficial to the public. . . . Each proprietor is entitled to such use of the stream, so far as it is reasonable, conformable to the uses and wants of the community, and having regard to the progress of improvements in hydraulics." *William Cary v. Albert Daniels* (Oct. 1844), MA Reports, Metcalf, 8:466.

27. *William Cowell v. Fisher Thayer* (Nov. 1842), MA Reports, Metcalf, 5:253.
28. *Robert Murdock v. Curtis Stickney* (Oct. 1851), MA Reports, Cushing, 8:113.
29. Ibid.
30. Ibid.
31. *Otis Howard and Another v. Proprietors of Locks and Dams on the Merrimack River* (Oct. 1853), MA Reports, Cushing, vol. 11. 267–268.
32. Ibid.
33. *The Palmer Company Petitioners v. Isaac Ferrill* (Sept. 1835), Pickering, 17:58.
34. *Spring v. Lowell.*
35. Jerome Smith, *Natural History of Fishes of Massachusetts, Embracing a Practical Essay on Angling* (Boston, 1833), 3.
36. Smith also edited the "Weekly News-letter" and published the *Medical Intelligencer*, the *Boston Medical and Surgical Journal*, the *Medical World*, and the standard textbook on anatomy.
37. Smith, *Natural History of Fishes*, 20.
38. Ibid., 134.
39. Ibid., 20.
40. Ibid., 344.
41. Ibid., 147.
42. Ibid., 344.
43. Ibid., 257.
44. Ibid., 257, 258.
45. Ibid., 169–170.
46. Thoreau, *Walden*, Henry David Thoreau, *Walden, or Life in the Woods* (Boston, 1854) 26.
47. In *Walden,* Thoreau offered advice to an Irish field hand, John Field. Leave the meadows in a wild state, Thoreau suggested to Field, and avoid hard work, and then Field would need fewer and lighter clothes. 174, 175, 176.
48. Ibid., 33.
49. See Lawrence Buell, *The Environmental Imagination: Thoreau, Nature Writing, and the Formation of American Culture* (Cambridge, 1995).

50. Thoreau, *A Week*, 40, 41.

51. Ibid. The farmers of Sudbury and Wayland not only stared at their flooded fields, they also took the Middlesex Canal Company to court. See *The Proprietors of the Sudbury Meadows v. The Proprietors of the Middlesex Canal* (Oct., 1839), MA Reports, Pickering, 23:36. See *William Heard v. Proprietors of the Middlesex Canal Company* (Oct. 1842), MA Reports, Metcalf, 5:81.

52. Thoreau, *Walden*, 177.

53. As the two brothers floated out of their town of Concord, the younger Thoreau reflected on the natural history of their surroundings. "Salmon, Shad, and Alewives were formerly abundant here . . . until the dam, and afterward the canal at Billerica, and the factories at Lowell, put an end to their migrations hitherward," he noted. Although Thoreau saw the "corporation with its dam" as the impediment to the fish, he could see no action to take: "Poor Shad! Where is thy redress?" Thoreau, the classic nineteenth-century antistatist, had no immediate answer. "Perchance, after a few thousands of years, if the fishes will be patient, and pass their summers elsewhere, meanwhile nature will have leveled the Billerica dam, and the Lowell factories and the Grass-ground River run clear again, to be explored by new migratory shoals." Thoreau, *A Week*, 35.

54. Transcendentalism's leading advocate was Thoreau's friend and patron Ralph Waldo Emerson. In the celebration of nature, the transcendentalists also believed they celebrated God, or a divine force. They also believed that the knowledge of the world was imperfect, and that to attain truth one needed to transcend one's assumptions. Contemplating nature would provide the means for such a transcendent state.

55. See Barbara Rosenkrantz, *Public Health and the State: Changing Views in Massachusetts, 1842–1936* (Cambridge, 1972), 10, 11, for an excellent discussion of the early efforts for public health in Massachusetts.

56. Ibid., 15–18.

57. Ibid., 21. Shattuck was influenced by the work done in Europe, particularly in Britain, in standardizing the names of diseases, and recording their incidences. He was in communication with the famous English sanitary reformer William Farr. See ibid., 24; see also Christopher Hamlin, *A Science of Impurity: Water Analysis in Nineteenth-century Britain* (Berkeley, 1990), for a discussion of Farr's influence in Great Britain.

58. The doctors were not sure why these smells caused disease. "These gasses are dangerous to health. What the specially noxious element in them is no one can define." MSBH, *First Annual Report of the State Board of Health, 1869* (Boston, 1870), 31. In 1842, Edwin Chadwick published *Sanitary Conditions of the Labouring Classes of Great Britain*, which led to widespread interest in the environmental causes of disease. Chadwick argued that the poor health conditions of the working classes were due to their poor environment, in particular to polluted water. See MSBH, *Fifth Annual Report of the State Board of Health, 1874* (Boston, 1875), 368, for a discussion of Chadwick's work. For a discussion of the emergence of anticontagionism among public health

activists, see Stanley Schultz, *Constructing Urban Culture: American Cities and City Planning, 1800–1920* (Philadelphia, 1989) 125–129; see also Rosenkrantz, *Public Health*, and Martin Melosi, *Garbage in the Cities: Refuse, Reform, and the Environment, 1889–1980*.

59. Rosenkrantz, *Public Health*, 26.
60. Ibid., 28.
61. Charles Chapin, *A Report on State Public Health Work* (Chicago, 1916).
61. CSBH, *First Annual Report of the State Board of Health* (Hartford, 1879), 37.
62. George Perkins Marsh, "Report, Made under Authority of the Legislature of Vermont on the Artificial Propagation of Fish" (Burlington, 1857), 8.
63. Ibid.
64. Ibid., 9.
65. Ibid., 13, 14.
66. Ibid., 15.
67. Ibid., 11.
68. Ibid., 12.
69. In *Man and Nature*, Marsh argued that modern technology did not cause the destruction of the natural world, but it did escalate destructive tendencies already inherent in humanity's desire to subdue the natural world. George Perkins Marsh, *Man and Nature, or Physical Geography as Modified by Human Action* (Cambridge, 1965).
70. Marsh, "Artificial Propagation of Fish," 13.
71. See Paul Johnson, *The Birth of the Modern: World Society, 1815–1830* (New York, 1991); and Christopher Lasch, *The True and Only Heaven: Progress and Its Critics* (New York, 1991).
72. Marsh, "Artificial Propagation of Fish," 11.
73. Ibid., 9.
74. Marsh, *Man and Nature*, 51, n. 53. Marsh goes on to note: "It is . . . the duty of government to provide all those public facilities . . . which are essential to the prosperity of civilized commonwealths" (51).
75. Marsh, "Artificial Propagation of Fish," 16.
76. Ibid., 16, 17.
77. For the story of Samuel Ely, see "Picturesque Hampden," quoted in Bennett, "Roots," 2:17.
78. *Edmund Baker v. The City of Boston* (Nov. 1831), MA Reports, Pickering, 11:183.
79. See Marsh, *Man and Nature*, 51, n. 53, for the suggestion that the answer to private wealth and privilege was the public agency of the state.
80. To say that radical changes in the environment led to a new, more proactive, role for the state is not to say that the state had previously been passive or laissez faire. Oscar and Mary Handlin argued almost fifty years ago that the state from the American Revolution onward had been actively promoting economic growth. Oscar Handlin and Mary Flug Handlin, *Commonwealth: A Study of the Role of Government in the American Economy, Massachusetts, 1774–1861* (Cambridge, 1961), 203–208.

CHAPTER 5

1. Lyman Diaries. The Lowell Proprietors through their Lake Company had built a dam on the lake and used the dam to regulate water flow to their mills downriver. Theodore Steinberg, *Nature Incorporated: Industrialization and the Waters of New England* (Cambridge, 1991), 135–165.

2. Smyth was active in and treasurer of the New Hampshire Agricultural Society and delivered several addresses on agriculture and farmers.

3. Lyman Diaries, Nov. 3, 1866, vol. 18.

4. Ibid., Dec. 30, June 26, 1865. These were not Lyman's only investments, but his investments in textile factories represented his greatest source of income. "Money flows in at a rate unprecedented." Lyman Diaries, Jan. 9, 1866, vol. 19. Lyman was on the board of directors of the Massachusetts Mills and the Atlantic Mills. Lyman Diaries, June 26, 1865, vol. 18; Jan. 8, 1877, vol. 35. For examples of his other holdings, see also Lyman Diaries, January 22, 1866, vol. 19; Apr. 27, Aug. 14, 18, Dec. 14, 1880, and June 15, 1881, vol. 38; Sept. 1, 1881, and Jan. 2, 1882, vol. 39; July 16, 1882, vol. 41. By the time Lyman graduated from Harvard, he was worth an estimated $400,000. For a biography of Lyman, see P.A.M. Taylor, "A New England Gentleman: Theodore Lyman II, 1833–1897," *Journal of American Studies* 17 (Fall 1983): 367–390. Lyman Scrapbook.

5. Lyman's letter to the *Chronicle*, Oct. 7, 1882, Lyman Scrapbook, vol. 47.

6. By the 1850s, the Boston Associates, a group of some fifteen families—the Lymans, the Lowells, the Cabots, the Quincys, the Eliots, the Blisses, the Lawrences, the Appletons, Edmund Dwight, Kirt Booth, Patrick Jackson, William Sturgis, Harrison Grey Otis, T. H. Perkins, and Israel Thorndike—effectively owned all of New England's textile mills. Richard W. Wilkie and Jack Tager, *Historical Atlas of Massachusetts* (Amherst, 1991), 30. This tight-knit group of families was also close socially. Theodore Lyman III was a friend and schoolmate of Nathan Appleton. He spent his summers with the Cabots at their place in Beverly. He dined with the Booths. See Lyman Diaries, vols. 14–44; for an example of the social links between these families, see May 15, 1880, vol. 38.

7. H. P. Bowditch, "Biographical Sketch of Theodore Lyman III, 1833–1897" (Washington, DC, 1903), manuscript, Museum of Comparative Zoology, Harvard University. See also Charles Francis Adams Jr., "Memoir of Theodore Lyman," Massachusetts Historical Society, *Proceedings* 20 (1906–1907), 147–177.

8. Nothing in either Lyman's early diaries or his later correspondence reflects any interest in the issue of slavery.

9. In 1882, Lyman, as a member of the Harvard Board of Overseers, voted against accepting a $50,000 gift for female musical education. Lyman Diaries, Jan. 11, 1882, vol. 40. He also opposed admission of women to the medical school. Lyman Diaries, Mar. 18, 29, Apr. 3, 12, 1882, vol. 40.

10. Lyman Scrapbook, vol. 46, "Farm Notebooks."

11. Lyman Diaries, June 15, 1883, vol. 41. On Dec. 14, 1880 Lyman Diaries, May 25, Dec. 4, 1880, and Oct. 12, 1889, vol. 38. Lyman also speculated in land in and around Boston and bought and sold waterpower coupons and railroad bonds. See, for example, Lyman Diaries, Mar. 26, 30, Apr. 1, 27, Aug. 18, 1880, vol. 38; Jan. 2, 1882, vol. 40.

12. For Lyman's extensive study of modern scientific improvementist agriculture, see Lyman Scrapbook, vol. 46, "Farm Notebooks."

13. See Tamara Plakins Thornton, *Cultivating Gentlemen: The Meaning of Country Life among the Boston Elite, 1785–1860* (New Haven, 1989), for a discussion of elite involvement in agricultural improvements. Note particularly the role of Lyman's grandfather and father in this movement.

14. Lyman Scrapbook, vol. 46. Lyman Diaries, May 14, June 9, Nov. 12, 1880, vol. 38; Oct. 13, 1882, vol. 40; July 13, 1883, vol. 41.

15. While in Paris, Lyman visited Comte and remained in constant correspondence with this leading philosopher of progressive positivism. Lyman Diaries, June 23, 1877, vol. 35. Lyman was a member and regular attendee of the Thursday Club, a club for scientifically interested intellectuals in Boston. On April 5, 1877, Lyman delivered a talk on "present scientific thought" that stressed positivism. Lyman Diaries, Apr. 5, 1877, vol. 35.

16. The Agassiz family looked upon Lyman as a son. He was a constant guest at their house. Lyman Diaries, Mar. 7, 28, Apr. 8, 1880, vol. 38. Lyman's closest friend was Agassiz's son Alexander. Lyman arranged for Alex to take over Harvard's Museum of Comparative Zoology. When winter set in, Lyman moved his family into an apartment in Alexander's Commonwealth Avenue home. While in Europe, Lyman met with the leading scientists of the various countries he visited. In France, he regularly worked at the Museum of Natural History. Lyman Diaries, Dec. 13–21, 1871, vol. 34. In January, Lyman saw Professor Ponzi in Rome, went to see several professors and scientists in Naples, and went back to Rome to dine with naturalist George Perkins Marsh. Lyman Diaries, Jan. 9, 18, Feb. 5, 17, 19, 26, 1872, vol. 35. Lyman then went on to see professors in Toledo and more in France. In Brussels, he was invited to a session at the Academy Circle and dinner with the Academy at the palace. Lyman then went on to visit scientists in Germany and Sweden. Lyman Diaries, Apr. 13, May 10, 14, 28, 29, 30, June 5, 6, July 1, 4, 7, 8, 11, Aug. 14, Sept. 11, 27, Oct. 17, 22, 25, 30, 1872, vol. 35.

17. See Lyman Diaries. On January 9, 1866, for example, forty members of the American Academy met at Lyman's house for a lecture on honeycombs and Darwin's theory of the making of the cell. Lyman Diaries, Jan. 9, 1866, vol. 19. A typical entry would be "Museum work as usual." Lyman Diaries, Apr. 2, 1880, vol. 38. In 1877, Lyman was elected treasurer of the American Academy of Arts and Science. Lyman Diaries, May 29, 1877, vol. 35. He was also a member of the American Association for the Advancement of Science. Lyman Diaries, Aug. 26, 1880.

18. Lyman Diaries, Jan. 10, 1877, vol. 35.

19. Lyman Diaries, Apr. 5, 1880, vol. 38; May 24, July 15, 1880, vol. 38; Apr. 18, 1881, vol. 39; Mar. 10, Apr. 19, May 4, 11, 1882, vol. 40; June 13, 1883, vol. 41.

20. Lyman Diaries, Oct. 11, 1882; Sept. 16, Sept. 24, Oct. 12, Nov. 4, Nov. 25, 1882, vol. 40. Letter to *Chronicle*, Oct. 7, 1882, Lyman Scrapbook, vol. 47. Lyman's concern about civil service reform was typical of that of many of his social class. These reform Republicans, known as mugwumps by their detractors, were typically upper-class elitists of conservative economic views.

21. Lyman Diaries, Oct. 12, 1882, vol. 40.

22. Lyman Diaries, Nov. 7, 1882, vol. 40.

23. Lyman Scrapbook, vol. 47.

24. Ibid. In Congress, Lyman supported civil service reform and worked for more government involvement in science. Lyman Diaries, Mar. 1, 8, June 28, May 3, 1884, vol. 42.

25. Ibid. Lyman Scrapbook, vol. 47.

26. Lyman Diaries, Apr. 26, 1884, vol. 42.

27. Among his other memberships, Lyman belonged to the Massachusetts Anglers' Association. Lyman Diaries, "note at end of 1875 Diaries," vol. 33.

28. Lyman Diaries, Oct. 16–19, 1866, vol. 22; Oct. 12, 13, 1880, vol. 38.

29. Lyman Diaries, Oct. 27, 28, 1879, vol. 35. When Lyman listed his qualifications to General Meade for his position as Meade's aide de camp, he listed his shooting ability gained from hunting. Taylor, "New England Gentleman," 379.

30. On August 9, 1865, only three months after his return from the army, Lyman and Louis Agassiz went cod fishing off the coast. In March of 1874, he persuaded his wife, Elizabeth, and Agassiz's son Alex to accompany him on a fishing trip. Lyman Diaries, Aug. 9, 1865, vol. 12; Mar. 2, 1874, vol. 32.

31. For examples of Lyman's fishing trips, see Lyman Diaries, Apr. 1, 8, 28, May 18, 26, 1874, vol. 32; Apr. 8, 29, May 4, 11, July 11, 1878, vol. 35; May 3, 14, June 5, 14, July 1, 1879, vol. 35; Mar. 31, Apr. 1, 7, 16, 26, May 5, 12, 13, 29, June 16, 17, July 13, 1880, vol. 38.

32. For Lyman's views on wilderness, see Lyman Diaries, Feb. 9, Mar. 8, 1880, vol. 35.

33. This is the man who said, "How I detest manufacturing towns." Lyman Diaries, June 28, 1867, vol. 22.

34. Lyman Diaries, Apr. 5, May 24, 1880, vol. 38; Apr. 18, 1881, vol. 39; Mar. 10, 18, Apr. 19, May 1, 11, 1882, vol. 40; June 13, 1883, vol. 41.

35. Morton Horwitz, *The Transformation of American Law, 1780–1860* (Cambridge, 1977). In 1818, the legislature passed an act "to authorize the Boston Manufacturing Company to shut the Fish Gate in their Dam across the Charles River," providing the company "shall make a fishway." MA Statute, Feb. 20, 1818. Then in 1823, the legislature in "An Act for the Relief of the Danvers Cotton Factory" enacted a statute that the Danvers Cotton Factory did not have to maintain a fishway. MA Statute, June 14, 1823. MCIF, *Sixth Annual Report, 1872*, 113, 146, 228.

36. *Proprietors of Sudbury Meadows v. The Proprietors of the Middlesex Canal Company* (Oct. 1839), MA Reports, Pickering, 23:36.

37. Comment made in 1820, quoted in Walter Hard, *The Connecticut* (New York, 1947).

38. *Marshall Fishing Company v. Hadley Falls Co.* (Sept. 1850), MA Reports, Cushing, 5:602.

39. Quoted in Marguerite Allis, *The Connecticut River* (New York, 1939), 126. Theodore Lyman felt that it was the dam at Millers River erected in 1798 that "doubtless stopped the salmon forever." Lyman Diaries, Oct. 12, 1865, vol. 18.

40. George Perkins Marsh was the first to suggest that deforestation was a major factor in the decline of New England's fish population. Deforestation affects fish reproduction by increasing erosion, causing silt to build up on riverbeds and ponds. The cutting of the forests disrupted the delicate ecosystem that the breeding streams and ponds needed. With the outflow of water from these breeding grounds, constant replenishments of nutrients are required. Lush beds of nitrogen-fixing lichen on the forest floor supply new sources of nutrients for the breeding grounds.

41. Allis, *Connecticut River*, 126; MCIF, *First Annual Report*, 40; George B. Goode, U.S. Commission of Fish and Fisheries, *The Fisheries and Fishery Industries of the United States* (Wash., D.C., 1887), 661. Timothy Dwight, *Travels Through New England and New York* (Cambridge, 1969), Volume 2, 213, 224. "Hadley," Judd Papers, 7:81. See also Lyman Papers.

42. Obstruction of the shad runs at Enfield by the locks and canals there, along with pollution in the water, so concerned members of an 1857 Massachusetts commission that they feared "they [shad] will soon leave the river entirely, unless something is done for their preservation." R. A. Chapman, "Artificial Propagation of Fish," *Report* (Boston, 1857), 10.

43. Lyman Hayes stated that early diaries and letters indicate the importance of salmon and shad for the early settlers in Vermont around Bellows Falls. The foot of Bellows Falls was the highest point of return for the shad, but twenty- and thirty-pound salmon continued over the falls. *The Connecticut River Valley in Southern Vermont and New Hampshire: Historical Sketches* (Rutland, 1929), 170, 171, 173.

44. The company initially attempted to avoid building the fishway. Steinberg, *Nature Incorporated*, 175.

45. Ibid.

46. In 1849, William McFarlin sued and won at the court of common pleas. The Essex Company appealed to the Massachusetts Supreme Judicial Court. It won its appeal because it was able to prove that McFarlin did not own the land from which he fished and could not prove the right by prescription since he could not show continuous fishing over twenty years. *William McFarlin v. Essex Company* (Oct. 1852), MA Reports, Cushing, 10:304.

47. MA Statute 1856, sec. 1, MCIF, *Sixth Annual Report*, 228.

48. Steinberg, *Nature Incorporated*, 180.

49. MA Statute 1856, MCIF, *Sixth Annual Report*, 228.

50. *Commonwealth v. Essex Company* (June 1859), MA Reports, Gray, 13:239.

51. Ibid.

52. Ibid.

53. See Horwitz, *Transformation of American Law.*

54. MA. Resolve 1856, MCIF, *Sixth Annual Report*, 228.

55. George Perkins Marsh, "Report, Made under Authority of the Legislature of Vermont on the Artificial Propagation of Fish" (Montpelier, 1857).

56. Chapman, "Artificial Propagation." Marsh, "Artificial Propagation of Fish," 16.

57. By the middle of the century, there were major fish-breeding and fish-stocking operations going in Europe. The Europeans, like their American counterparts, argued that fish breeding represented a science that would allow for the return of nature's abundance. J. W. Patterson, "An Address Delivered before the Game and Fish League of New Hampshire, April 2, 1878," NHCF, *Report of the Commissioners on Fisheries, 1878* (Concord, 1878), 54.

58. The *New England Farmer*, a widely read farm journal of the midcentury, was continually singing the virtues of scientific farming, which would lead to improved wealth and better health. See Richard Judd, *Common Lands, Common People: The Origins of Conservation in Northern New England* (Cambridge, 1997), and Margaret Richards Pabst, *Agricultural Trends in the Connecticut Valley Region of Massachusetts, 1800–1900*, Smith College Studies in History, vol. 26, nos. 1–4 (Northampton, Oct. 1940–July 1941), for a discussion of scientific farming. I would like to thank Richard Judd for providing me with examples of the interest in scientific agriculture from the *New England Farmer.*

59. NHCF Report, 1878, 52, 57.

60. Chapman, "Artificial Propagation," 14. George Perkins Marsh concluded that artificial propagation of fish was "thus far without important results economical or physical." Marsh, "Artificial Propagation of Fish," 18, 19.

61. Chapman, "Artificial Propagation," 4. The committee felt that although their experiments failed, the success of pisciculture elsewhere proved that artificial propagation of fish was not only possible but profitable. "The increase of our population . . . will be likely to give [the preservation and improvement of the fisheries] new importance." Ibid., Chapman, "Artificial Propagation," 8, 9. Agassiz testified before a joint committee of the Massachusetts Legislature in 1865 that he had limited success raising trout in the washbasin of his bedroom. JSC, "Joint Special Committee Report," MA Senate, No. 183 (Boston, Apr. 1865), 6 (hereafter JSC, "Report on the Obstruction").

62. Ibid., 10.

63. Ibid., 12. Like Marsh, the Massachusetts commissioners believed that successful fish breeding should be carried on as a private enterprise with state support. Marsh, "Artificial Propagation of Fish," 17.

64. Chapman, "Artificial Propagation," 12, 13.

65. The Massachusetts commissioners argued that the legislature should pass laws giving proprietors on large bodies of water the rights to incorporate and be chartered to establish fisheries and have exclusive rights to the fish. Ibid., 12.

66. NHCF, Report *Made to the Legislature of New Hampshire, 1866* (Concord, 1866), 8 (hereafter, NHCF Report, 1866).

67. Ibid., 4. "Resolutions of the State of Vermont," quoted in JSC, "Report on the Obstruction," 2. Lyman claimed that it was Judge Henry Bellows of New Hampshire who initiated New Hampshire's law, but the judge was not alone. New Hampshire's governor at the time, Joseph Gilmore, also complained that Massachusetts was not doing enough to protect the fish. Lyman Papers, vol. 48. New Hampshire also demanded attention be paid to the impact of industrial wastes on fish. NHCF Report, 1866, 4.

68. JSC, "Report on the Obstruction," 3, 4.

69. Ibid., 4.

70. Charles Storrow from the Essex Company testified that the mills needed all the water, and that the fishways that were built got washed out three times in floods. Speaking for the mills at Holyoke, Stewart Chase claimed that it was not practicable to construct permanent fishways over the Holyoke dam. Ibid., 17, 19.

71. Louis Agassiz spoke before the committee and stressed the importance of pollution, "chiefly [from] manufacturing establishments . . . acids, dyestuffs and all sorts of mixtures, and still more by the building of saw-mills," on declining fish numbers. Sawmills were a problem because they destroyed the spawning grounds. Agassiz did note that if the obstacles could be removed and the waters cleaned up, New England's waters, with the aid of fish breeding, could be restored to their earlier abundance. Ibid., 5, 6.

72. Ibid., 21, 22, 23. The committee felt that New Hampshire was harmed by the dam at Lawrence, and that Vermont, which was not even consulted, might have grounds for complaint about the Holyoke dam. At the hearings, Francis, representing the interests of the manufacturers, argued against a fishway, but he did allow the possibility that this problem might "be surmounted by a sufficient expenditure of money" (21). It was this possibility that the committee focused on in its final report.

73. Ibid., 22.

74. Ibid.

75. Massachusetts felt that it had "made careful provision for the maintenance and protection of the fishing rights of the citizens of New Hampshire, . . . but . . . it must be conceded that an injury to some extent has been inflicted upon the rights of New Hampshire." Ibid., 21, 22.

76. MA Resolve, May 1865, MCIF, *Sixth Annual Report*, 246.

77. Ibid.

78. The difficulty of this task was clear. New Hampshire was adamant that the "only obvious obstructions to fish migrations are the dams at Lawrence on the Merrimack and Holyoke on the Connecticut," at both of which the manufacturers were just as adamant that they were not going to build fishways. NHCF Report, 1866, 9, 10.

79. The commissioners noted that not only had salmon stopped running up river, but even the numbers of shad had significantly declined after the 1820s. From the 1840s to the 1850s, the number of Connecticut River shad

caught in Massachusetts dropped 20 percent, and between 1853 and 1865, it dropped another 8 percent. COPF, *Report of the Commissioners, 1866,* (Boston, 1866), 34, 35, 36, 37.

80. Ibid., 9.

81. Ibid., 13. The commissioners noted that "plainly, then, the supply of water at Lawrence is abundant in the spring" (11).

82. Ibid., 12.

83. For a discussion of the interest in pollution and fish breeding, see Donald J. Pisani, "Fish Culture and the Dawn of Concern over Water Pollution in the United States," *Environmental Review* 8 (1984): 117–130.

84. COPF, "Report of the Commissioners," 14.

85. Ibid., 14, 15.

86. Ibid., 19.

87. Ibid., 21.

88. Ibid., 16. In 1866, the Massachusetts fish commissioners had the water at Lawrence tested by Hayes and S. Dana Hayes to determine its ability to sustain fish life. The Hayeses claimed that "a stream of impure water flowing into the river does not soon enter the general current, but is confined to one side and the bottom and before becoming diffused the noxious matter is deposited or decomposed and, in short, it soon works itself clear." Quoted in NHCF Report, 1866, 11.

89. COPF, "Report of the Commissioners," 40, 41.

90. Ibid., 51.

91. See NHCF Report, 1878, for New Hampshire's fisheries laws; in 1865, New Hampshire outlawed dams or obstructions on the Connecticut and Merrimack (14).

92. NHCF Report, 1866, 3, 4.

93. Vermont's first fish commissioners were Albert Hayes and Charles Barrett from the Connecticut River Valley.

94. Lyman Diaries, Mar. 28, 1866, vol. 19. The New Hampshire commissioners with whom Lyman met were Judge H. Bellows and Captain W. Sandborn. Vermont's commissioner at the meeting was Professor A. Hager. Vermont Commission on Fisheries, *Report of the Fish Commissioners of Vermont, 1869* (Montpelier, 1869), 7.

95. MCIF, *Sixth Annual Report, 1872,* 250, 251.

96. Lyman Diaries, July 3, 1866, vol. 19. Massachusetts's other commissioner was Alfred Field. Lyman Diaries, July 12, 1866, Vol. 19.

97. Ibid., July 12, 31, 1866. Despite this initial cooperation, the fish commissioners in New Hampshire were critical of the fishways developed in Massachusetts, and tensions continued between the two states. In 1870, the New Hampshire commissioners reported that they believed that the fishways at Lowell, Lawrence, and Manchester were "unsatisfactory." NHCF, *Report Made to the Legislature of New Hampshire, 1870* (Concord, 1870), 3. A year later, the New Hampshire commissioners reported that "the views of the Massachusetts commissioners do not exactly coincide with ours in relations to the constructions of fishways." *Report Made to the Legislature of New Hamp-*

shire, 1871 (Concord, 1871), 3. For further objections, see *Report Made to the Legislature of New Hampshire, 1873* (Concord, 1873), 4.

98. Commissioners of Fisheries, *Report of the Commissioners of Fisheries* (Boston, 1867), 2, 3. Lyman Diaries, Sept. 8, Oct. 2, Oct.16, 1866.

99. Lyman Diaries, Feb. 25, 1867, vol. 22.

100. Ibid., Nov. 15, 1867, vol. 23. Vermont Commissioners on Fishery, *Report of the Fish Commissioners of Vermont, 1867* (Montpelier, 1867).

101. Lyman Diaries, Dec. 19, 1867, and Feb. 19, 1868, vol. 23; Oct. 5, 1876, vol. 34; Commissioners of Fisheries, *Fourth Annual Report of the Commissioners of Fisheries, 1870*, Sen. Doc. 12 (Boston, 1871), 8.

102. Lyman Diaries, July 12, 31, 1866.

103. Lyman Diaries, Sept. 18, 1867.

104. MCIF, *Eighth Annual Report*, Sen. Doc. 49 (Boston, 1874), 18, 26.

105. MCIF, *Second Annual Report*, House Doc. 60 (Boston, 1868), 19.

106. NHCF Report, 1878, 49.

107. Ibid.

108. Ibid., 48, 49, 61.

109. NHCF Report, 1866, 15.

110. NHCF Report, 1871, 4, 5.

111. Ibid., 4.

112. Theodore Lyman, "Remarks before the Massachusetts Joint Committee on Fisheries," Mar. 11, 1869 (Boston, 1869), 11.

113. MCIF, *Third Annual Report*, 64.

114. Lyman, "Remarks," 1.

115. MCIF, *Fourth Annual Report*, Sen. Doc. 12 (Boston, 1870), 42.

116. MCIF, *Third Annual Report*, Sen. Doc. 3 (Boston, 1869), 18.

117. Lyman, "Remarks," 15.

118. In this sense, Lyman shared the vision of resources argued by Garrett Hardin in "The Tragedy of the Commons," *Science*, 162:3859 (Dec., 1968) 1243–1248.

119. MCIF, *Second Annual Report*, House Doc. 60 (Boston, 1868), 21. Lyman's hostility to the republicanism he attributed to Tisdale's neighbors reflects his belief that protection of a resource could not be entrusted to the public exercising its common law rights.

120. Commonwealth of Massachusetts, "An Act for Encouraging the Cultivation of Useful Fishes," Senate Doc. 91 (Boston, 1866). New Hampshire and Vermont fish commissions were also working on the fish breeding project. NHCF Report, 1878, 21; Vermont Commission on Fishery, *Report of the Fish Commissioners of Vermont, 1867* (Montpelier, 1867), 8, 9.

121. Lyman, "Remarks," 11.

122. MCIF, *Fifth Annual Report*, Sen. Doc. 11 (Boston, 1871), 33.

123. Lyman Diaries, Apr. 13, May 14, 1868, vol. 23. New Hampshire also had problems with some manufacturers. After promising to build a fishway, the Franklin Falls Company failed to build adequate fishways, forcing New Hampshire to take the company to court. NHCF Report, 1869, 10, and 1871, 6.

124. MCIF, *Fifth Annual Report*, 37.

125. Ibid., 38.

126. *Commissioners on Inland Fisheries v. Holyoke Waterpower Company* (Mar., 1870), MA Reports, Browne, 8:446. The commissioners won in the court of common pleas, but the Holyoke Water Power Company appealed to the Supreme Judicial Court. Judge Gray, writing for the Massachusetts court, ruled that "the rights of the public are . . . not to be presumed to have been surrendered to a corporation, except so far as an intention to surrender them clearly appears in the charter. The grant of a franchise from the Commonwealth for one public object is not to be unnecessarily interpreted to the disparagement of another." "The right to have migratory fish pass, in their accustomed course up and down rivers and streams . . . is also a public right, and may be regulated and protected by the legislature in such a manner, through such commissioners . . . as it may deem appropriate." It was this ruling that the Holyoke Company appealed to the federal courts.

127. MCIF, *Eighth Annual Report*, 62.

128. Lyman Diaries, Oct. 5, 6, 1876, vol. 34.

129. The fishways around the dams at Lowell and Lawrence proved a partial success. Yet salmon numbers remained small, and the shad population continued to be problematic. Lyman blamed overfishing, but a more likely cause was the continued pollution of the river and damage to the breeding grounds, along with the inadequacy of the fishways. Lyman Diaries, Sept. 8, Oct. 2, 3, 1866, vol. 18; Apr. 1, 1876, vol. 34; Sept. 28, 1877, vol. 34. For a discussion of overfishing, see Lyman Diaries, Nov. 11, 1876, vol. 34; July 16, 1878, vol. 35; July 27, Aug. 21, 1880, vol. 38.

130. MCIF, *Seventh Annual Report*, Sen. Doc. 8 (Boston, 1873), 17.

131. CCFG, *First Biennial Report, 1895–1896* (Hartford, 1896), 10.

132. Vermont Commission on Fishery, *Report of the Fish Commissioners of Vermont, 1867* (Montpelier, 1867), 6.

CHAPTER 6

1. The law creating the Massachusetts board stated that "the board shall take cognizance of the interests of health and life among the citizens of the Commonwealth. They shall make sanitary investigations and inquiries in respect to the people, the causes of disease, and especially epidemics, the sources of mortality and the effects of localities, employments, conditions, and circumstances on the public health." MSBH, *First Annual Report of the State Board of Health, 1869* (Boston, 1870), 6. See also Henry I. Bowditch, *Public Hygiene in America* (Boston, 1877), and Bowditch, "Origins," 6. *Bowditch Correspondence*, 2:217. The two volumes of *Bowditch Correspondence* contain Bowditch's correspondence to his family and friends as well as his journal entries.

2. Bowditch's father, Nathan Bowditch, was a world-famous mathematician and a much published author. *Bowditch Correspondence*, 1:2. Bowditch sat on several medical boards and on the Board of Trustees of the Boston Public Library and was a member of the famous intellectual group the Thursday

Club. Bowditch to John Whittier, Dec. 16, 1887, *Bowditch Correspondence*, 2:320; Bowditch Papers.

3. *Bowditch Correspondence*, 2:7.

4. Bowditch, "Origins," 2–5.

5. Barbara Rosenkrantz, *Public Health and the State: Changing Views in Massachusetts, 1842–1936* (Cambridge, 1972).

6. *Bowditch Correspondence*, 2:218. Lemuel Shattuck had also advocated a state board in the 1850s. Rosenkrantz, *Public Health*, 46, 47.

7. George Hoyt Bigelow, "Henry P. Walcott," *New England Journal of Medicine* 207, 23 (December 1932): 1002.

8. Bowditch "Origins," 3. Before the Massachusetts Board of Health, there had been city boards. See Rosenkrantz, *Public Health*, 10, 11, 12.

9. Rosenkrantz also noted that the bill creating the state board of health passed at a time in which both Democrats and Republicans were generally supportive of state regulatory boards. In 1855, Louisiana had a state board of health, predating Massachusetts's, but Louisiana's board was really the board of health only for New Orleans. Charles Chapin, *A Report on State Public Health Work* (Chicago, 1916), 61.

10. In a letter to the Reverend Brazer of Salem, June 27, 1835, Bowditch stated, "I am a radical!" *Bowditch Correspondence*, 1:105, 114.

11. Reminiscences about Bowditch by Dr. William Henry Thayer, *Bowditch Correspondence*, 1:146.

12. *Bowditch Correspondence*, 1:273–279. Bowditch complained that the Boston elite was indifferent to slavery and "ostracized him"; other doctors warned Bowditch that his view would destroy his career. *Bowditch Correspondence*, 1:100.

13. Ibid., 1:206. Bowditch's views also included antidiscrimination. When the Massachusetts General Hospital Board of Directors passed a rule in 1841 excluding "coloreds," Bowditch resigned. Bowditch's protest was effective. The hospital rescinded its exclusionary rule. Letter, May 28, 1841, *Bowditch Correspondence*, 1:130.

14. Ibid., 2:182.

15. Journal, May 9, 1867, *Bowditch Correspondence*, 2:212. In 1882, Bowditch objected to Harvard overseers' refusal to accept women into the medical school. Lyman Diaries, Nov. 3, 1866, Mar. 28, Apr. 3, 12, 1882, vol. 40.

16. *Bowditch Correspondence*, 1:112, 128.

17. On the way home from work October 21, 1835, Bowditch came upon a crowd trying to get to Garrison and Thompson for "holding a meeting in opposition to slavery." Bowditch was aghast: "It is time for me to become an abolitionist!" *Bowditch Correspondence*, 1:99, 100.

18. In a letter to his parents, Bowditch noted that because of his father's connections, "everybody is telling me of the great advantages I have over any other American." Bowditch to his parents, June 29, 1832, *Bowditch Correspondence*, 1:20. For his friendship and involvement with Holmes, Greene, Warren, and Jackson, see Bowditch to his parents, Nov. 17, 1832, *Bowditch Correspondence*, 1:32, and Bowditch to his sister, Dec. 17, 1832, *Bowditch Correspondence*, 1:33, and Jan. 6, 1833, *Bowditch Correspondence*, 1:35, 36.

19. Bowditch to his mother, Jan. 27, 1833, *Bowditch Correspondence*, 1:37.

20. Bowditch to his parents, Mar. 13, 1833, *Bowditch Correspondence*, 1:41, 42.

21. Bowditch to his mother, Feb. 13, 1834, *Bowditch Correspondence*, 1:73.

22. Bowditch to his father, May 13, 1833, *Bowditch Correspondence*, 1:50.

23. Bowditch to his father, July 31, 1834, *Bowditch Correspondence*, 1:86, and undated, *Bowditch Correspondence*, 1:89. Henry Bowditch's father Nathaniel to Olivia, Mar. 10, 1838, *Bowditch Correspondence*, 1:95, 96.

24. Bowditch's opposition to slavery led him to increasingly view his social equals with suspicion. Although some of Boston's elite, like Bowditch's friend Cabot and the Quaker Poet, John Whittier, were active in the radical abolitionist cause, Bowditch felt that "all the great men of respectability stood aloof." Journal, Aug. 11, 1842, *Bowditch Correspondence*, 1:162. Bowditch's involvement in radical causes also linked him to a variety of European social activists. He was a personal friend of Harriet Martineau and was one of the very few people she would see as her health failed. Henry Bowditch to his wife, July 17, 1859, *Bowditch Correspondence*, 1:323.

25. See Philip Terrie, *Contested Terrain: A New History of Nature and People in the Adirondacks* (Syracuse, 1997).

26. Journal, Aug. 25, 1866, *Bowditch Correspondence*, 2:69, 70. For an example of Bowditch's feelings about the wilderness of the Adirondacks, see Bowditch to his wife, July 1865, *Bowditch Correspondence*, 2:61–66. In 1869, Bowditch visited Dr. Shattuck at his cottage on Chateaugay Lake. Journal, Aug. 10, 1869, *Bowditch Correspondence*, 2:66.

27. Journal, June 26, 1849, *Bowditch Correspondence*, 1:192–195.

28. Journal, Oct. 31, 1846, *Bowditch Correspondence*, 1:189–190.

29. Bowditch's radicalism combined American republican ideology and its attacks against privilege with nineteenth-century French egalitarianism.

30. In 1858, Derby had delivered an address to the Boston Social Science Association arguing for an active state agency that would intervene to prevent ill health. Rosenkrantz, *Public Health*, 51.

31. John Hoadley, an engineer on the board, had also been among the radical abolitionists.

32. MSBH, *Fifth Annual Report of the State Board of Health, 1873* (Boston, 1874), 369. See Reginald Reynolds, *Cleanliness and Godliness* (New York, 1974), for a discussion about the ideology of cleanliness.

33. Charles Rosenberg and Carrol Smith-Rosenberg, "Pietism and the Origins of the American Public Health Movement: A Note on John H. Griscom and Robert M. Hartley," in Judith Leavitt and Ronald Numbers, eds., *Sickness and Health in America: Readings in the History of Medicine and Public Health* (Madison, 1978), 345–358. For a discussion of the public health movement and its concern for environmental causes of disease, see John Duffy, *The Sanitarians: A History of Public Health* (Urbana, 1990).

34. Health reformers tested the water by walking downriver and looking for clearness or turbidity, and live minnows. CSBH, *Tenth Annual Report of the State Board of Health* (Hartford, 1888), 215.

35. Ibid., 252. This report contained a description of the water.

36. Ibid., 219, 312.

37. NHSBH, *First Annual Report* (Concord, 1882), 1:54.

38. As the New Hampshire Board of Health noted in 1882, "water rapidly purifies itself" by mixing with other water and running "a few miles." NHSBH, *First Annual Report*, 1:14. The common belief at the time was that "water became pure by exhalation and oxygenation in running four miles from the point where the sewage is received" (246). In 1880, the Connecticut State Board of Health found that the sewage dumped into the Little or Park River was not a problem because the river had enough flow to cleanse itself by the time it reached Hartford. CSBH, *Second Annual Report of the State Board of Health* (Hartford, 1880), 19. As late as 1895, the Connecticut State Board of Health argued that the Park River sewage from New Britain had a change in the content of its sewage by "dilution with clean water from the watershed." CSBH, *Eighteenth Annual Report of the State Board of Health, 1895* (Hartford, 1896), 227. In 1891, the Connecticut State Board of Health, although concerned about the health impact of drinking Connecticut River water, reported that "the sewage entering the Connecticut River is so largely diluted as to have a scarcely perceptible effect on the chemical composition of the water as a whole." CSBH, *Fourteenth Annual Report of the State Board of Health, 1891* (Hartford, 1892), 439. For discussion of the germ theory, see John Cumbler, "Whatever Happened to Industrial Waste: Reform, Compromise, and Science in Nineteenth-Century Southern New England," *Journal of Social History* 29, 1 (Fall 1995): 149–171. For a discussion of anticontagionism, see Nancy Tomes, "The Private Side of Public Health: Sanitary Science, Domestic Hygiene, and the Germ Theory, 1870–1900," *Bulletin of History of Medicine*, Winter 1990, vol. 64, 4, 309–631; Stanley Schultz, *Constructing Urban Culture: American Cities and City Planning 1800–1920* (Philadelphia, 1989), 125–129; see also Rosenkrantz, *Public Health*.

39. MSBH, *Fourth Annual Report of the State Board of Health, 1872* (Boston, 1873), 40.

40. CSBH, *Second Annual Report*, 73.

41. MSBH, *Ninth Annual Report of the State Board of Health, 1877* (Boston, 1878), 103.

42. CSBH, *Tenth Annual Report*, 261.

43. NHSBH, *Third Annual Report* (Concord, 1884), 3: 297.

44. Hartford built just such a system, moving sewage out of the city directly to the Connecticut River. As the Connecticut became polluted along Hartford's waterfront, the system was expanded to push the sewage farther out into the flow of the river. CSBH, *Twenty-second Annual Report of the State Board of Health, 1899* (Hartford, 1900), 283. New Hampshire's board of health found that "the first thing ... do[ne] with a polluted water system is to lay a main line parallel with the main watercourse thus intercepting the polluted drainage that would otherwise gather in the watercourse." NHSBH, *Third Annual Report*, 3:297, 300.

45. MSBH, *Fourth Annual Report*, 109, 110.

46. The Connecticut Board of Health noted in 1896 that "the increasing contamination of certain of our rivers shows [the need for] some other method of sewage disposal than the discharge of crude sewage into rivers." CSBH, *Eighteenth Annual Report*, 201.

47. MSBH, *Fourth Annual Report*, 40.

48. MSBH, *Eighth Annual Report of the State Board of Health, 1876* (Boston, 1877), 406.

49. Lynn Margolin and William Sedgwick, *The Principles of Sanitary Science and the Public Health* (New York, 1918).

50. MSBH, *Eighth Annual Report*, 207, 209.

51. CSBH, *Fourteenth Annual Report*, 438.

52. CSBH, *Tenth Annual Report*, 40, 41. Conditions downstream from New Britain got so bad that in 1897, a Mr. Nolan sued the city of New Britain for dumping sewage into Piper's Brook. The court ruled for Nolan, noting that New Britain's sewage made "the condition of the stream so impure that it could not be used for domestic purposes, and for the watering of cattle, and so that it gave off noxious and unhealthy odors." *Nolan v. City of New Britain* (1897), 38 A. 703, CT Reports 69, 668.

53. CSBH, *Tenth Annual Report*, 40, 41.

54. See *Massachusetts Reports*, (Boston) 1804–1880. See particularly *Rufus Weston v. Elijah Alden* (Oct. 1811), MA Reports, Tyng, 8:137, *Hinsdale Smith and Another v. The Agawan Canal Company* (Sept. 1861), MA Reports, Allen, 2:355, and *City of Springfield v. Samuel Harris* (Sept. 1862), MA Reports, Allen, 4:494, for examples of the courts defining water rights in terms of quantity. For examples of the courts seeing water rights in terms of quality, see *William Merrifield v. Nathan Lombard* (Oct. 1866), MA Reports, Allen, 8:16, *William Merrifield v. City of Worcester*, MA Reports, Browne, 14:216, and *Troy Cotton and Woollen Manufacturing Company v. City of Fall River* (Feb. 1883), MA Reports, Lathrop, 20:267.

55. Benjamin Silliman, *Remarks Made on a Short Tour Between Hartford and Quebec in the Autumn of 1819* (New Haven, 1824), 10, 11.

56. Letter to the Editor, *Springfield Union*, in "Parks and Pollution" file, Connecticut River Valley Historical Society, Springfield, Massachusetts.

57. See *James Shermin v. Fall River Iron Works Company* (Oct. 1861), MA Reports, Allen, 2:524.

58. *Royal Call v. Otis Allen* (Jan. 1861), MA Reports, Allen, 1:137.

59. *Royal Call v. Otis Allen*.

60. CSBH, *Second Annual Report*, 77.

61. See James Olcott's speech before the Agricultural Board of Connecticut, CSBH, *Ninth Annual Report of the State Board of Health* (Hartford, 1887), 239, 241, 242. MSBH, *First Annual Report of the State Board of Health, 1869* (Boston, 1870), 15.

62. MSBH, *Fourth Annual Report*, 40.

63. CSBH, *Second Annual Report*, 25.

64. CSBH, *Seventeenth Annual Report of the State Board of Health* (Hartford, 1895), 224.
65. Bowditch, "Origins," 25.
66. Ibid., 26.
67. Ibid., 15.
68. Ibid., 11.
69. Ibid., 14. Bowditch to W. H. Hunt, Nov. 13, 1887, *Bowditch Correspondence*, 231; Journal, Oct. 18, 1842, *Bowditch Correspondence*, 114.
70. Bowditch to his daughter, undated, *Bowditch Correspondence*, 228.
71. Bowditch, "Origins," 8–9.
72. Ibid., 14.
73. MSBH, *First Annual Report*, 15.
74. George Fredrickson, *The Inner Civil War: Northern Intellectuals and the Crisis of the Union* (New York, 1965).
75. Bowditch, "Origins," 11.
76. Bowditch Papers, vol. 3, no. 3; Journal, May 26, 30, 31, 1870, *Bowditch Correspondence*, 149–199.
77. Bowditch to his daughter, July 25, 1870, *Bowditch Correspondence*, 153, 171.
78. Ibid., 171.
79. Ibid., 174.
80. Bowditch to his wife, Sept. 24, 1870, *Bowditch Correspondence*, 180–181.
81. Bowditch to the Massachusetts Board of Health, Dec. 30, 1870, Bowditch Papers, vol. 3, no. 3. The English had recognized stream pollution by midcentury. In 1865, a royal commission on river pollution was appointed to, among other things, determine "how far by new arrangements, the refuse arising from industrial processes can be kept out of the streams or rendered harmless," and a second commission was appointed in 1868. In 1876, the British Pollution Prevention Act was passed. This act prohibited "the discharge into any streams of any poisonous, noxious, or polluting liquids." Quoted in Phelps, "Stream Pollution," 198.
82. Although Bowditch was excited about the reform movement he found in England, he was also aware of its limitations. In its 1873 report, the Massachusetts board noted that "unless legislation to prevent pollution of streams can be better enforced in Massachusetts than in England, we may conclude that the spoiling of our rivers as sources of water-supply is a question of time of density of populations and of their size." MSBH, *Fourth Annual Report*, 100.
83. Bowditch, "Origins," 9.
84. MSBH, *Fifth Annual Report*, 354.
85. In its first report, the Connecticut board noted that "when the Commonwealth charged this board to take cognizance of the best interests of the health and life of its citizens, it simply expressed a broader recognition of that accepted duty which has heretofore led it to pass and enforce laws for the protection and advancement of their moral, intellectual, and material benefit, as now for their physical." CSBH, *First Annual Report of the State Board of Health* (Hartford, 1879), 48.

86. In 1879, Connecticut spelled out the duties of its Board of Health as, among others, "to prevent the pollution of air and water by foul liquids, gasses, vapors, and dirt of all kinds." Ibid., 39.

87. MSBH, *First Annual Report*, 16.

88. NHSBH, *First Annual Report*, 1:6.

89. NHSBH, *Third Annual Report*, 3:207, 108.

90. Bowditch, "Origins," 9.

91. MSBH, *Fifth Annual Report*, 364.

92. CSBH, *Tenth Annual Report*, 176.

93. MSBH, *Fifth Annual Report*, 369.

94. NHSBH, *First Annual Report*, 1:5, 18.

95. MSBH, *First Annual Report*, 15, 16.

96. NHSBH, *First Annual Report*, 1:23.

97. NHSBH, *Fifth Annual Report of the New Hampshire Board of Health* (Concord, 1886), 5:13.

98. MSBH, *Fifth Annual Report*, 49.

99. Ibid.

100. NHSBH, *First Annual Report*, 1:212, 213.

101. Ibid., 213.

102. Ibid., 254, 255.

103. CSBH, *Fourth Annual Report of the State Board of Health* (Hartford, 1882), 28.

104. MSBH, *Third Annual Report of the State Board of Health, 1871* (Boston, 1872). William Ripley Nichols and George Derby, "Sewerage, Sewage: The Pollution of Streams: The Water Supply of Towns," MSBH, *Fourth Annual Report 1872* (Boston, 1873).

105. MSBH, *Fourth Annual Report*, 20.

106. Ibid., 4–8, 10, 11, 20; MSBH, *Fifth Annual Report*, 63–150; James Kirkwood, William Ripley Nichols, and Frederick Winsor, "The Pollution of Rivers: An Evaluation of the Water-Basins of the Blackstone, Charles, Tauton, Neponsit, and Chicopee Rivers, with General Observations on Water Supplies and Sewerage," MSBH, *Seventh Annual Report, 1875* (Boston, 1876).

107. MSBH, *Ninth Annual Report*.

108. Ibid., 71.

109. Ibid., 73–79. Despite its prohibition against industrial pollution, the board did recognize the needs of industrialists: "It would be unwise to place too great restrictions upon manufacturers setting up, for all cases, some arbitrary standard of purity, which must be always followed, but which could not be enforced."

110. MSBH, *Eighteenth Annual Report, 1886* (Boston, 1887), 278.

111. In 1884, the Connecticut board noted that "there seems to be a need of further legislation upon this subject [pollution of streams] especially when the water is so contaminated by manufacturing wash and sewage as to be detrimental to health.... Sewage and wastes from factories and manufacturing establishments make up the polluting materials." CSBH, *Fourth Annual Report*, 28.

112. MSBH, *Eighteenth Annual Report*, 278.

CHAPTER 7

1. CSBH, *Ninth Annual Report of the State Board of Health* (Hartford, 1887), 241.
2. MSBH, *Fourth Annual Report of the State Board of Health*, 1872 (Boston, 1873), 21.
3. *William Merrifield v. Nathan Lombard* (Oct. 1866), MA Reports, Allen, 13:16.
4. *William Merrifield v. City of Worcester* (Oct. 1872), MA Reports, Browne, 14:216.
5. CSHB, *Ninth Annual Report of the State Board of Health* (Hartford, 1887), 241.
6. Bowditch, "Origins," 32–36.
7. Ibid.; *Bowditch Correspondence*, 226.
8. After the passage of the antipollution act, attempts by the board of health to force several manufacturers to stop polluting were met with resistance by manufacturers. See Bowditch, "Origins," 43; see also Barbara Rosenkrantz, *Public Health and the State: Changing Views in Massachusetts, 1842–1936* (Cambridge, 1972), 87.
9. Bowditch Papers, 8:245.
10. Bowditch, "Origins," 36.
11. Bowditch Papers, 4:31.
12. When the merger with the boards of lunacy and charity occurred, the old board became the health committee of the new board.
13. Bowditch, "Origins," 43.
14. Katherine E. Conway and Mabel Ward Cameron, *Charles Francis Donnelly: A Memoir with an Account of the Hearings on the Bill for the Inspection of Private Schools in Massachusetts in 1888–1889* (New York, privately printed, 1909) (hereafter, Donnelly Memoir).
15. Donnelly was also president of the Charitable Irish Society. In his work for Bishop Williams, Donnelly represented the church's position against the attempt by conservative Protestants to regulate Catholic schools. Ibid., 32.
16. Contrary to Bowditch's claim, Donnelly was not a manufacturer, but his family had interests in textile mills in southern New England. Ibid., 237.
17. Ibid., 22, 23, 32. As the chair of the board of lunacy and charity, Donnelly recommended tightening up restrictions on getting relief (235).
18. Bowditch, "Origins," 36; Donnelly Memoir, 22, 23. Shortly after the boards were merged, Benjamin Butler was elected governor.
19. Butler's campaign for cleanliness during his occupation of New Orleans won high praise among sanitarians. Rosenkrantz, *Public Health*, 49.
20. George Hoyt Bigelow, "Henry P. Walcott" *New England Journal of Medicine* 207, 23 (Dec. 1932): 1001–1002. Rosenkrantz, Public Health, 91–92.
21. Walcott was on the board of overseers at Harvard, was a fellow of the corporation, and was appointed to the presidency of the college in Charles Eliot's absence. He was a member of the American Academy of Arts and Sciences and served on the board of Massachusetts General Hospital, ulti-

mately becoming chairman. Like Lyman, Walcott was active in promoting parks and belonged to several citizen groups concerned over public parks. In 1886, he was president of the American Public Health Association. Walcott was also elected president of the American Academy of Arts and Sciences. He was an honorary fellow of the Royal Sanitary Institute of Great Britain. He became the chairman of the Metropolitan Water and Sewerage Board and a member of the first Metropolitan Drainage Commission. Walcott was also the president of the Fifteenth International Congress of Hygiene and Demography. *Proceedings of the American Academy of Arts and Sciences* 68, 13 (Dec. 1933): 687–688. Rosenkrantz, Public Health, 92. "Henry P. Walcott" *New England Journal of Medicine.*

22. "Henry P. Walcott" *New England Journal of Medicine.*

23. Judith Rosenkrantz implies that Walcott was a moderate reformer reluctantly drawn into battle over public health issues. Nonetheless, his involvement in public health, initially sparked by Shattuck's work, soon led him to an ever more radical position. By the end of the 1880s, he was arguing not only that the poor should not be subjected to ill health because of their poverty, but that they had a right to a clean and healthy environment, and it was the responsibility of the state to guarantee that right.

24. Massachusetts was not the only New England state to confront the problem of industrial pollution and face resistance from manufacturers. The Connecticut Board of Health, established in 1879, also faced resistance from manufacturers. "In the case of streams, it is vastly more difficult [to protect purity because] the manufacturing interests are affected." CSBH, *Second Annual Report of the State Board of Health* (Hartford, 1880).

25. Quoted in Rosenkrantz, "Public Health," 83.

26. *Inhabitants of Brookline v. Charles Mackintosh* (July 1882), MA Reports, Lathrop, 19:215.

27. Ibid.

28. Robinson was educated at Harvard and in 1873 with the support of local Republicans and businessmen ran for public office, first in the state legislature, then in Congress.

29. Bowditch, "Origins," 43.

30. Donnelly Memoir, 240; Rosenkrantz, "Public Health," 78. Although the conflict between manufacturers and pollution reformers was most intense in New England, the most industrialized region of the country, other areas and other countries also experienced this conflict. See Earle Phelps, "Stream Pollution by Industrial Wastes and Its Control," in M. Ravenel, ed., *A Half Century of Public Health* (New York, 1921), 202–207. See also Joel Tarr, "Industrial Wastes and Public Health: Some Historical Notes, Part 1, 1876–1932," *American Journal of Public Health* 75 (1985): 1059–1067.

31. Bowditch, "Origins," 43.

32. Ibid., 45.

33. Walcott argued that Donnelly wanted his position on the board to protect manufacturers' interests. The conflict between Walcott and Donnelly also involved a question of patronage. See Rosenkrantz, Public Health, 90–91.

34. Ibid., 85.

35. The *Advertiser* joined in Robinson's defense and claimed the *Globe* was only interested in the story as a means of attacking Republicans.

36. Bowditch, "Origins," 46.

37. Quoted by Bowditch in "Origins," 48. For the series of stories about Walcott's removal, see *Boston Herald*, Dec. 8–17, 1885.

38. Bowditch, "Origins," 49, 50. Donnelly Memoir, 23, 24. Donnelly remained the chair of the board of lunacy and charity (24, 27). After the board was recreated as an independent board in 1886, Walcott remained chair for the next twenty-eight years.

39. "An Act to Protect the Purity of Inland Waters," Acts and Resolves 1886, c 230.

40. Ibid. Henry P. Walcott came on in 1882 with the support of many of the original members. Walcott took over as chair of the health committee in 1882. Walcott was removed as chair and board member by Governor Robinson in 1885. Although Walcott was reappointed to the board, his battle to defend a public health position may have engendered a certain amount of caution. See Rosenkrantz, Public Health, 88, for a discussion of Walcott's cautious approach as president of the American Public Health Association.

41. See Theodore Steinberg, *Nature Incorporated: Industrialization and the Waters of New England* (Cambridge, 1991), 235–239, for a biography of Mills and a discussion of his involvement in pollution reform.

42. MSBH, *Fifth Annual Report, State Board of Health, 1873* (Boston, 1874), 355. Bowditch argued that progress was measured in the improved health of the citizens, and if the state had an interest in progress, then the state had a vital interest in promoting public health. "Is there room here in this field of human life [public health] for governmental cooperation?" (363). Using the example of education, Bowditch argued that the state had a direct interest in educating its population for the general betterment, and it should likewise work to protect the general health of the population. Bowditch pointed to a long history of State of Massachusetts involvement in pubic health."The law offers some protection against contagious diseases . . . also against nuisances. . . . In some degree it regulates tenement houses; it endeavors to save children from the exhaustive effect of over-labor in factories" (363, 364).

43. Ibid., 373.

44. Ibid., 369.

45. Donnelly Memoir, 32.

46. Ibid., 24.

47. See Thoreau's Journal, 12:387 Henry David Thoreau, "Journal" in *Works*, 12:387. See also Lawrence Buell, *The Environmental Imagination: Thoreau, Nature Writing, and the Formation of American Culture* (Cambridge, 1995), 136, 212, 213, and Donald Worster, *Nature's Economy: A History of Ecological Ideas* (Cambridge, 1985).

48. See Donald Pisani, "Promotion and Regulation: Constitutionalist and the American Economy," *Journal of American History* 74 (Dec. 1987): 740–768.

49. As the New Hampshire Commissioners of Fisheries stated, "the rights of the whole community" could no longer be disregarded by the mill owners. NHCF, *Report Made to the Legislature of New Hampshire, 1873* (Concord, 1873), 3.

50. MSBH, Fifth Annual Report, 356, 357.

51. George Fredrickson in *The Inner Civil War: Northern Intellectuals and The Crisis of the Union* (New York, 1965) argues that the postwar reformers looked to the state as a new agent for social reform. In the postwar period, reformers tended to be divided over the role of the state and that of business. What shape that reform would take and how the state should function divided the reformers between moderates and radicals. Certainly two of our reformers, Bowditch and Lyman, fit Fredrickson's model. Lyman, a conservative reformer, believed that the state could bring together business and the public through the agency of science and technology. Bowditch, a radical, believed the state should be an instrument to control business.

52. Lyman believed that water pollution was simply wasteful. COPF, *Report of the Commissioners, 1866* (Boston, 1866), 21.

53. *Bowditch Correspondence*, 221.

54. MSBH, *Fifth Annual Report*, 361.

55. See Fredrickson, *The Inner Civil War*, for a discussion of the role of science and professionalism in postwar reform activities, 180–122.

56. See Steinberg, *Nature Incorporated*, for a discussion of this quest.

57. *Elizabeth W. Emerson v. Lowell Gas and Light Company* (Jan. 1863), MA Reports, Allen, 6:46. The gas company's expert, Prof. Eben Horsford, was accepted by the court as having expertise because he had studied the influence of gas on health.

58. Fredrickson argued that the war experience motivated postwar reformers to see the state differently.

59. Although I think that Fredrickson is correct in noting the importance of the war in shaping postwar understandings of the role of the state, it is important not to overemphasize the change that emerged. Except for the brief Reconstruction experiment, it wasn't until the end of the century that we get any kind of government activity, and that activity was modeled more on the actions of states such as Massachusetts.

60. See MSBH, *Fifth Annual Report*, 364, 365, 372, 374, for examples of this transatlantic flow of information.

CHAPTER 8

1. CSBH, *Twenty-seventh Annual Report of the State Board of Health*, (Hartford, 1905), 4.

2. Ibid., 56.

3. CSBH, *Twenty-eighth Annual Report of the State Board of Health*, (Hartford, 1906), 8.

4. For reformers' concern over child labor and women's work, see Edith Abbott, *Women in Industry: A Study in American Economic History* (New York,

1910); Jane Addams, *A New Conscience and an Ancient Evil* (New York, 1913), and *Twenty Years at Hull House* (New York, 1910); Katherine Anthony, *Mothers Who Must Earn* (New York, 1914); Elizabeth Faulkner Baker, *Protective Labor Legislation* (New York, 1925); Louis Brandeis, *Women in Industry* (New York, 1908); Helen Campbell, *Prisoners of Poverty: Women Wage-Workers, Their Trades and Their Lives* (Boston, 1887), and *Women Wage-Earners: Their Past, Their Present, Their Future* (Boston, 1893). For concern over industrial diseases, see Alice Hamilton, *Industrial Poisons in the United States* (New York, 1935), and *Exploring the Dangerous Trades: The Autobiography of Alice Hamilton, M.D.* (Boston, 1943). For reformers' concern over tenement house conditions, see Jacob Riis, *How the Other Half Lives* (New York, 1890).

5. The Connecticut board noted in 1910 that in Massachusetts, "where these same problems arose some years earlier," the state had passed antipollution laws. CSHB, *Thirty-First Annual Report of the State Board of Health* (Hartford, 1911), 10.

6. CSHB, *Eighth Annual Report of the State Board of Health, 1885* (Hartford, 1886), 345–351; NHSBH, *First Annual Report*, vol. 1 (Concord, 1882); Charles Caverly and Henry Tinkham, "History of the Medical Profession in Vermont," in Walter Hill Crockett, ed., *Vermont–The Green Mountain State* (New York, 1923), 5:624.

7. The New Hampshire Board of Health in its first report to the state noted that "the right of all persons to pure air, unpolluted water, and uncontaminated soil is a natural inheritance. This right is protected and maintained by law." Also, the board of health had the right to "suppress any ordinary unhealthy nuisance." NHSBH, *First Annual Report*, 1:233, 239.

8. Robinson also appointed two other doctors who pleased the medical society and the Boston Society for Medical Improvement. The two manufacturers Robinson picked were from the Connecticut River Valley. Hiram F. Mills was chosen to represent technological expertise, but as the chief engineer of the Essex Company in Lawrence, he also represented industry. With the exception of Walcott, none of the board members had served on the earlier boards. The board was also authorized and funded to hire additional full-time workers, including a physician as medical officer and secretary to the board, and three engineers to supervise reports on water and sewage of the towns and cities of the state.

9. Quoted in Barbara Rosenkrantz, *Public Health and the State: Changing Views in Massachusetts, 1842–1936* (Cambridge, 1972), 87.

10. In 1905, the Connecticut Board of Health reported that "the safeguarding of public health" required the board to "discover and meet the enemy [the polluter] and defeat him." CSBH, *Twenty-eighth Annual Report*, 8.

11. In 1893, Mills became a consultant for the Proprietors of Locks and Canals at Lawrence.

12. Acts and Resolves 1886, "An Act to Protect the Purity of Inland Waters," C274, 230. Rosenkrantz, *Public Health*, 86, 87; Theodore Steinberg, *Nature Incorporated: Industrialization and the Waters of New England* (Cambridge, 1991), 235–239.

13. See Thomas Brock, ed., *Milestones in Microbiology* (Englewood Cliffs, 1965), 65–120, for examples of articles by these researchers on their discovery of the germ theory. See also M. Ravenel, ed., *A Half Century of Public Health* (New York, 1921), 13–16; see also Frederick Gorham, "A History of Bacteriology and Its Contribution to Public Health Work," 66–93, in *A Half Century of Public Health*, and Nancy Tomes, "The Private Side of Public Health: Sanitary Science, Domestic Hygiene, and the Germ Theory, 1870–1900," *Bulletin of History of Medicine*, Winter 1990, 509–631. See CSBH, *Seventh Annual Report of the State Board of Health* (Hartford, 1885), 158–167, for a discussion of this work and its increasing significance for American public health officials.

14. MSBH, *Eighth Annual Report of the State Board of Health, 1876* (Boston, 1877), 118.

15. In 1885, the Connecticut State Board of Health noted that Hartford suffered a diphtheria epidemic that killed 200 people. The Board of Health believed that the cause of the epidemic was the open sewer of the Park River, which receives wastes from 43,000 people. "The effluvia from [the river] which fill and impregnate the air" carried the disease. "The relationship between incomplete sewerage, tainted air and disease is by no means new." CSHB, *Seventh Annual Report*, 170, 175.

16. CSBH, *Third Annual Report of the State Board of Health* (Hartford, 1881), 6. Even as late as 1890, the Connecticut board was arguing that infant diarrhea was caused by polluted air. "It is now fairly well established that diarrheal diseases are caused by inhalations of poisoned air." CSHB, *Twelfth Annual Report of the State Board of Health* (Hartford, 1890), 29.

17. NHSBH, *Second Annual Report* (Concord, 1883), 2:60.

18. NHSBH, *Sixth Annual Report* (Concord, 1887), 6:182.

19. NHSBH, *First Annual Report*, 1:75. We can also see this ambivalence in Connecticut. In its 1885 report, the Connecticut State Board of Health extensively discussed the work of microbiologists in Europe and the germ theory, yet in that same year the board noted that "slightly diluted sewage . . . produces 'excrement-reeking' air which . . . produces most poisonous effects." CSBH, *Seventh Annual Report*, 176.

20. By 1885, the Connecticut board was willing to accept the germ theory for most diseases. "The microscope with its improvements . . . coupled with new and approved methods of investigations, have given to those minute bodies a legitimate standing as etiological factors of the utmost importance in disease." CSBH, *Seventh Annual Report*, 157. CSBH, *Twenty-sixth Annual Report of the State Board of Health* (Hartford, 1904), 221. For a discussion of the resistance of Americans to the germ theory, see Phyllis Richmond, "American Attitudes toward the Germ Theory," *Journal of History of Medicine* vol. 9, 3, 1954, 428–454.

21. CSBH, *Seventh Annual Report*, 157.

22. CSBH, *Eighth Annual Report*, 228.

23. CSBH, *Fourteenth Annual Report of the State Board of Health* (Hartford, 1892), 437.

24. MSBH, *Report upon the Metropolitan Water Supply* (Boston, 1895), 188. See Christopher Hamlin, *A Science of Impurity: Water Analysis in Nineteenth Century Britain* (Berkeley, 1990), for a discussion of the role of chemists in water analysis.

25. See Robert Clark, *Ellen Swallow: The Woman Who Founded Ecology* (Chicago, 1973), for a discussion of Swallow's reform activity.

26. See MSBH, *Twentieth-eighth Annual Report of the State Board of Health, 1896* (Boston, 1897), for a description and detailed analysis of the state's water supply and river water quality.

27. MSBH, *Twentieth Annual Report of the State Board of Health, 1888* (Boston, 1889), x, xi.

28. Ibid., 32, 33. See Clarke, *Ellen Swallow*, 147–148, for Swallow's influence on the development of scientific sanitation reform.

29. By the second year, the board was spending $7,792 at Lawrence and $1,200 at MIT. Rosenkrantz, *Public Health*, 100.

30. See Joel Tarr, *The Search for the Ultimate Sink: Urban Pollution in Historical Perspective* (Akron, 1996), 293–308, for a discussion of using urban wastes for fertilizer.

31. NHSBH, *Sixth Annual Report*, 6:190.

32. Ibid.

33. See Hamlin, *Science of Impurity*, for a discussion of English efforts.

34. Steinberg, *Nature Incorporated*, 233. Typhoid (Eberthella typhi), like cholera and dysentery, is spread by a salmonella gram-negative bacilli or bacteria being ingested through the mouth. It is most easily spread through drinking water contaminated by excrement or urine of those already infected. The typhoid bacilli can live in water for long periods of time and pass alive great distances through water. Those who contact typhoid suffer from fever, a rose-colored eruption, abdominal pains, splenomegaly, diarrhea, and dehydration.

35. MSBH, *Twentieth-eighth Annual Report*, 190–194, 204–207; MSBH, *Twentieth-second Annual Report of the State Board of Health, 1890* (Boston, 1891), 533.

36. MSBH, *Twentieth-third Annual Report of the State Board of Health, 1891* (Boston, 1892), 668–705.

37. CSBH, *Fourteenth Annual Report*, 439. CSBH, *Fifteenth Annual Report of the State Board of Health* (Hartford, 1893), 205.

38. MSBH, *Twentieth Annual Report*, 32, 33.

39. For a summary of the impact of the Massachusetts studies and experimental station, see CSBH, *Seventeenth Annual Report of the State Board of Health* (Hartford, 1895), 224. As early as 1882, the New Hampshire board noted that "too much reliance is often placed upon the idea that water rapidly purifies itself. . . . The fallacy of such a doctrine becomes more and more apparent." NHSBH, *First Annual Report*, 1:14.

40. CSBH, *Seventeenth Annual Report*, 224.

41. For a discussion of this debate, see Tarr, *Search for the Ultimate Sink*, 131–158.

42. Charles Chapin, *A Report on State Public Health Work* (Chicago, 1916), 48. For a discussion of the emergence of the scientific expert in public

health reform, see Rosenkrantz, *Public Health*, 75, 76, 82, and Joel Tarr, "Disputes over Water Quality Policy: Professional Cultures in Conflict, 1900–1917," *Journal of Public Health* 70 (Apr. 1980): 427–435. See also Joel Tarr, "Searching for a 'Sink' for an Industrial Waste: Iron-making Fuels and the Environment," *Environmental History Review* 18 (Spring 1994) 9–34; Craig Colten, "Creating a Toxic Landscape: Chemical Waste Disposal Policy and Practice, 1900–1960," *Environmental History Review* 18 (Spring 1994): 85–116; Christopher Sellers, "Factory as Environment: Industrial Hygiene, Professional Collaboration, and Modern Sciences of Pollution," *Environmental History Review* 18 (Spring 1994) 55–84.

43. CSBH, *Eighth Annual Report*, 229.

44. Ibid.

45. MSBH, *Twenty-eighth Annual Report*, 81–405.

46. CSBH, *Twenty-second Annual Report of the State Board of Health* (Hartford, 1900), xxiv. By 1915, the Connecticut health officials were using dissolved oxygen as the measure of water purity. CSBH, *Thirty-fourth Report of the State Board of Health* (Hartford, 1917), 169.

47. Caverly and Tinkham, "Medical Profession in Vermont," 625.

48. CSBH, *Eighteenth Annual Report of the State Board of Health* (Hartford, 1896), 201.

49. For recommendations of the Connecticut Board of Health, see CSBH, *Seventeenth Annual Report*, 221–224; xvii, 290; *Twentieth Annual Report of the State Board of Health* (Hartford, 1898), 268; *Twenty-second Annual Report*, xvii. In 1901, the Connecticut board reported that "the question of sewage purification is one which is being forced most prominently on the attention of many of our cities." CSBH, *Twenty-fourth Annual Report of the State Board of Health* (New Haven, 1902), 316. For litigation against upstream polluters, see CSBH, *Twentieth Annual Report of the State Board of Health, 1897* (Hartford, 1898), 325, 326; *Morgan v. Danbury*, *Nolan v. City of New Britain*, 38 A. 703, 69 CT 668. In both of these cases, the Connecticut Supreme Court supported an injunction against the cities for discharging pollution into the waters. The new technology of filtration systems raised the issue suggested in *Merrifield v. City of Worcester* that an upstream polluter could be found guilty of a nuisance against a downstream riparian owner, if the upstream user failed to maintain or use existing technology. See Christine Rosen, "Differing Perceptions of Value of Pollution Abatement across Time and Place: Balancing Doctrine in Pollution Nuisance Law, 1840–1906," *Law and History Review* 2 (Fall 1993), 303–381, for a discussion of the court's approach to pollution and nuisance.

50. CSBH, *Twenty-second Annual Report*, xvii. CSBH, *Nineteenth Annual Report of the State Board of Health* (Hartford, 1897), 290. By 1899, New Britain, in response to lawsuits, was buying land in Berlin to construct a system to purify its sewage, and South Manchester was building a sewage filtration system to cut the wastes that it was dumping into the Hockhanum River. CSBH, *Twenty-third Annual Report of the State Board of Health* (New Haven, 1901), 21, 334.

51. CSBH, *Twentieth Annual Report*, 268.
52. Ibid., 333.
53. CSBH, *Tenth Annual Report of the State Board of Health* (Hartford, 1888), 196.
54. Ibid., 203–215.
55. CSBH, *Seventh Annual Report*, 177.
56. Ibid., 183.
57. Ibid., 188. One of the industrial wastes the board felt did little harm was sulphate of copper, which although doing "little harm" was by the board's own admission "the most poisonous of any substance known to [fish]" (188).
58. Given the science of the time, no one connected the health of fish to morbidity of humans from industrial poisons ingested in small amounts over long periods.
59. CSBH, *Seventh Annual Report*, 183. The board did note that "it was desirable, whenever possible that the waters of our streams may be kept sufficiently pure to permit the growth of fish." But given a choice between fish and an industry whose pollutants did not contain germs, and hence in the board's view was not dangerous to humans, the board felt it had to support industry (183).
60. The *Boston Herald*, Sept. 28, 1896, noted that "under the direction of the State Board of Health the struggle for the amelioration of certain unhealthy . . . conditions existing for half a century has finally begun."
61. Tarr, "Disputes over Water Quality Policy," 431–432. Tarr argued that public health professionals divided over the issue of cleaning up wastes before dumping or filtering the water before using. He finds that while the old scientific generalists, particularly doctors, wanted to clean wastes before dumping, the new scientific specialists, particularly sanitary engineers, argued that a cost-benefit analysis suggested filtering the water before use would be more efficient. See also Joel Tarr, James McCurley, and Terry Yosie, "The Development and Impact of Urban Wastewater Technology: The Changing Concepts of Water Quality Control, 1850–1930," in Martin Melosi, *Pollution and Reform in American Cities* (Austin, 1980), 59–82.
62. For a discussion of the dual system verses the single system, see Tarr, *Search for the Ultimate Sink*.
63. See CSBH, *Nineteenth Annual Report*, 321, 322, for complaints about municipalities dumping wastes into local streams.
64. CSBH, Twentieth Annual Report, 326.
65. Ibid.
66. CSBH, *Twenty-first Annual Report of the State Board of Health* (Hartford, 1899), 242.
67. CSBH, *Fourteenth Annual Report*, 437.
68. MSBH, *Thirty-Fourth Annual Report of the State Board of Health, 1902* (Boston, 1903), 313.
69. Ibid., 313, 314.
70. Ibid.

71. Ibid., 312.

72. CSBH, *Thirty-first Annual Report*, 10. In the same report, the Connecticut State Board of Health suggested a law similar to the Massachusetts law that would "prohibit the depositing of sewage, factory wastes, or any polluted matter into streams or lakes" (10, 11).

73. Ibid., 334. In the same report, the Board of Health noted that Springfield, although it did not treat its sewage, did have pipes that extended far enough out into the Connecticut to reduce the nuisance of the sewage (334). In the committee's special report of 1909, the investigator for the Hartford Joint Committee assumed the extension of the sewer line from the Franklin Ave District was an acceptable solution to its sewage problem because it "now discharges into the Connecticut River at a point much further out from the shore where the water is deep and the current swift, so that for several years at least I can not see how this sewage can be offensive or create a nuisance." Hartford Department of Engineering, *Report of the Joint Special Committee of the Court of Common Council on Flood Control, Sewage Facilities, and Sewage Disposal* (Hartford, 1909), 78.

74. Ibid., 373, 374, 375.

75. See CSBH, *Twentieth Annual Report*, 308, for a discussion of Hartford's intercepting system.

76. MSBH, *Thirty-Fifth Annual Report of the State Board of Health, 1903* (Boston, 1904), 97.

77. Hartford Department of Engineering, *Report of the Joint Special Committee*, 39.

78. Ibid., 47.

79. Ibid., 78.

80. Ibid., 47.

81. MSBH, *Thirty-Fourth Annual Report*, 377.

82. Ibid.

83. CSBH, *Twenty-eighth Annual Report*, 81.

84. MSBH, *Fortieth Annual Report of the State Board of Health, 1908* (Boston, 1909), 549, 550. The smaller tributaries to the Connecticut that received sewage directly from towns such as Northampton on the Mill River were "badly polluted." The Deerfield River became "very offensive" due to the increase of sewage dumped into its waters. The French River was "seriously polluted" by manufacturing wastes (from woolen mills in Rockdale, North Oxford, and Oxford) and sewage from the town of Webster. The Hoosick River was polluted by sewage from Adams, North Adams, and Williamstown, besides large quantities of manufacturing wastes.

85. Earle Phelps, "Stream Pollution by Industrial Wastes and Its Control," in M. Ravenel, ed., *A Half Century of Public Health* (New York, 1921), 206–207.

86. CSBH, *Third Annual Report*, 139. In the same report, the board worried that as the city increased in size, the value and necessity of parks would be even greater, "one of the inevitable gifts to the future."

87. By the twentieth century, public health advocates were focusing more on issues such as school inspection and inoculations. See CSBH, *Twenty-fourth*

Annual Report, 264; *Twenty-seventh Annual Report*, 80, 81; MSBH, *Thirty-ninth Annual Report of the State Board of Health, 1907* (Boston, 1908), 548.

88. In 1898, the Connecticut Board of Health noted that "the question of purification of surface waters is largely a financial matter.... One phase of the question, affecting particularly the financial side, is the wasteful use of water common to American communities. It is the uniform experience that the use of meters eliminates a large part of the waste among consumers." CSBH, *Twenty-second Annual Report*, 241. Focusing on cleaning up germs in drinking water also had potential problems. Death rates for dysentery, typhoid, and cholera were declining in New England, but reports of deaths by cancer were climbing, although these figures are highly questionable. MSBH, *Twenty-Ninth Annual Report*, xxxvii. In Connecticut, cancer deaths per 10,000 rose from 3.25 in 1879 to 8.25 in 1909. CSBH, *Thirty-first Annual Report*, 51, chart 14. See also CSBH, *Twenty-seventh Annual Report*, 43, chart 12.

89. MSBH, *Thirty-fourth Annual Report*, 315. In their report, despite its concern about industrial wastes, in the end the board had little it could recommend to the state other than to suggest that sewer lines be significantly submerged and empty into large enough bodies of water so as to prevent these industrial wastes from accumulating on rivers and stream edges or pooling near neighborhoods.

CHAPTER 9

1. The Connecticut State Board of Health noted in 1921 that despite attempts to deal with sewage, public works had not been able to keep up with rapid urban and industrial growth. Industrial Wastes Board, Second Biennial Report, 1920, *First and Second Biennial Reports of the Industrial Wastes Board 1916–1920* (Hartford, 1921), appendix, 2 (hereafter, IWB, Second Biennal Report).

2. See Joel Tarr, "Searching for a 'Sink' for an Industrial Waste: Ironmaking Fuels and the Environment," *Environmental History Review* 18 (Spring 1994) 9–34; and Craig Colten, "A Historical Perspective on Industrial Wastes and Groundwater Contamination," *Geographical Review* 81 (Apr. 1991) 215–228, and *Industrial Wastes in the Calumet Areas, 1869–1979* (Champaign, 1985).

3. MSBH, *Twenty-Eighth Annual Report of the State Board of Health, 1896* (Boston, 1896, 428. The shift of concern away from industrial wastes also occurred on the national level. Earle Phelps, "Stream Pollution by Industrial Wastes and Its Control," in M. Ravenel, ed., *A Half Century of Public Health* (New York, 1921), 197–208.

4. Earle Phelps in a report to the Public Health Service in the second decade of the twentieth century noted that the presence of pollution, including industrial wastes, "dulls the aesthetic sense of a community." Quoted in Joel Tarr, *The Search for the Ultimate Sink:Urban Pollution in Historical Perspective* (Akron, 1996), 367.

5. In 1909, the Connecticut Commission of Fisheries and Game noted a "steady decline of shad . . . [as the] river has become little more than an open

sewer." CCFG, *Eighth Biennial Report, 1909–1910* (Hartford, 1910), 8. The *Hartford Daily Courant* noted on February 3, 1913, that prohibiting the pollution of the state's rivers and streams was necessary to maintain fish stock, particularly shad. Also see the *Hartford Daily Courant* editorial, May 23, 1913, "A Lesson from Fishes," which argued that pollution was driving away the fish. The substance of the editorial was that water that was too polluted for fish was also too polluted for people. For the concern of the region's oyster fishers, see *Hartford Daily Courant*, Jan. 25, 1913.

6. CCFG, *Sixth Biennial Report, 1905–1906* (Hartford, 1906), 20. *Hartford Daily Courant*, Feb. 3, 1913.

7. *Hartford Daily Courant*, Feb. 3, 1913.

8. CCFG, *Eighth Biennial Report*, 29.

9. CCFG, *Ninth Biennial Report, 1911–1912* (Hartford, 1912), 25.

10. *Hartford Daily Courant*, Feb. 3, 1913.

11. Ibid., Jan. 29, 1913; Feb. 3, 1913.

12. Ibid., Jan. 20, 1913.

13. CSBH, *Report on the Investigation of the Pollution of Streams* (Hartford, 1915), 6. *Hartford Daily Courant*, Feb. 6, 1913.

14. CSBH, *Investigation of the Pollution of Streams*, 6.

15. Ibid.

16. Ibid.

17. CSBH, *Thirty-third Report of the State Board of Health* (Hartford, 1915), 11.

18. Focusing on bacterial problems, the board erroneously believed that "damage to human health from eating fish from polluted streams is . . . slight." Ibid.

19. The board did note that polluted water was a health concern for oysters, clams, and other shellfish. Ibid.

20. CSBH, *Investigation of the Pollution of Streams*, 6.

21. Phelps was trained at MIT and was an assistant bacteriologist at the Lawrence Experimental Station from 1899 to 1900. See Tarr, *Search for the Ultimate Sink*, 366, for a discussion of the development of DO as a measure of water quality.

22. The BOD created an "oxygen sag" where the DO dropped after exposure to wastes. In 1913, the Public Health Service began an intensive study of water pollution at its Center for Pollution Studies in Cincinnati, employing Phelps. See ibid., 365–366, for a discussion of the study.

23. Health reformers had realized the role of oxygen in oxidizing pollutants since the 1870s. "The readily oxidizable part of the sewage that gains access to a river is destroyed in the few first miles run." CSBH, *First Annual Report of the State Board of Health* (Hartford, 1879), 89.

24. Ibid., 7.

25. Ibid., 8.

26. Ibid., 11.

27. CSBH, *Investigation of the Pollution of Streams*, 12.

28. Investigators looked for DO levels above 6 ppm as signs of relatively healthy rivers.

29. CSBH, *Investigation of the Pollution of Streams*, 24, 35.

30. Ibid., 19.

31. See *Edward Haskell v. City of New Bedford* (Oct. 1871), MA Reports, Browne, 12:208, for the court's opinion that one of the natural purposes of running water was to dispose of wastes.

32. See *Hartford Daily Courant*, for the month of June 1913, for advertisements of "bathing suits for men and women."

33. See ibid., Aug. 19, 23, 1913. On August 16, 1913, Kathyn Cepiano, a bookkeeper, drowned while swimming in the Connecticut despite the fact that she was known to regularly swim in the river. *Hartford Daily Courant*, Aug. 19, 1913.

34. Samuel Hays has noted that increased concern for recreation spurred a renewed activism around environmental issues in the post-war period. What I am suggesting here is that the concern for recreation originated earlier, but it was revitalized in the post-war period. See Samuel Hays, *Beauty, Health, and Permanence: Environmental Politics in the United States, 1955–1985* (Cambridge, Eng., 1987). Family swimming and canoeing outings were mostly confined to middle-class or working-class urban residents. *Hartford Daily Courant*, Aug. 25, 1913.

35. Ibid., Aug. 7, 1913.

36. Ibid., Jan. 3, 1913.

37. In 1915, reformers just barely failed to get the Massachusetts legislature to enact a bill that would have prohibited the pollution of streams by substances that made them "poisonous or dangerous to fish or animal life" or vegetation. Tarr, *Search for the Ultimate Sink*, 359.

38. IWB, Second Biennial Report, 28.

39. Ibid., 2, 3. New Britain built its system in response to law suits.

40. Ibid.

41. CCFG, *Fifth Biennial Report, 1903–1904* (Hartford, 1904), 19. See the *Hartford Courant*, Feb. 3, 1913, on how a series of bills had been proposed to curb industrial pollution but how manufacturers had managed to either defeat the bills or amend them in such a way as to exempt industrial pollution. See also CCFG, *Ninth Biennial Report*, 25.

42. *Hartford Daily Courant*, Jan. 25, Feb. 6, Aug. 23, 1913.

43. Ibid., Feb. 3, Jan. 29, 1913.

44. Ibid., May 9, 1913; see May 23, 1913, for an editorial in favor of anti-pollution legislation.

45. Ibid., Apr. 17, 1913.

46. Ibid.

47. Ibid., March 8, 18, 1913.

48. Ibid., April 23, 1913.

49. For examples of this, see the debate over the state's working men's compensation bill.

50. *Hartford Daily Courant*, June 6, 1913.

51. For a discussion of the lobbying efforts of the National Manufacturers Association, see ibid., June 30, 1913.

52. Connecticut Statute of 1915, ch. 284 and 306.
53. CSBH, *Thirty-sixth Annual Report of the State Board of Health* (Hartford, 1920), 253.
54. Ibid., 212.
55. CSBH, *Tenth Annual Report of the State Board of Health* (Hartford, 1888).
56. CSBH, *Thirty-Sixth Annual Report*, 253.
57. Ibid.
58. Ibid.
59. Ibid., 213.
60. CCFG, *Ninth Biennial Report*, 25.
61. Ibid., 238.
62. It was better, Phelps believed, to wait until "experimental progress" provided a profitable way to dispose of pollution. Phelps, "Stream Pollution," 206–207. The question of whether to force manufacturers to clean up pollution before technology can support a profitable alternative or to wait until a profitable alternative is developed remains with us today.
63. The Industrial Wastes Board was created by the legislature as a part of the Board of Health. Their reports were published in the state Board of Health Reports. CSBH, *Thirty-Sixth Annual Report*, 214.
64. Ibid., 215, 216. In its investigation of the Hockanum River, the industrial wastes board argued that the "logical method of purifying the river is to treat sewage and industrial wastes at or near the source of pollution." It also argued that separate treatment of the industrial wastes will be necessary." Ibid., 238, 239.
65. Ibid., 215, 216.
66. Ibid., 226.
67. Ibid., 227.
68. Ibid.
69. Ibid., 257.
70. Ibid., 258.
71. Ibid., 264.
72. Ibid., 265.
73. Tarr, *Search for the Ultimate Sink*, 360.
74. William Leuchtenburg, *Flood Control Politics: The Connecticut River Valley Problem, 1927–1950* (Cambridge, 1953), 25.
75. CSBH, *Twenty-eighth Annual Report of the State Board of Health* (Hartford, 1906), 2.
76. Ibid., 2, 3.
77. Ibid., 4.
78. CSBH, *Twenty-seventh Annual Report of the State Board of Health* (Hartford, 1905), 4.
79. Ibid., 7, 8.
80. CSBH, *Thirty-sixth Annual Report*, 264.
81. In 1918, the board noted that manufacturers on the Naugatuck River were willing to work with the board, if the board would accept the leadership of the manufacturers association already working on the issue. The

manufacturers' preferred solution was "dilution," which the board found unacceptable. The board suggested instead that the manufacturers treat their wastes. The manufacturers did not find this proposal acceptable. IWB, Second Biennial Report, 28.

82. See Tarr, *Search for the Ultimate Sink.*

CHAPTER 10

1. Edward Bellamy, "Notebook, Eliot Carson," Bellamy Papers.
2. Edward Bellamy, manuscript, AM, 1181.3, Binder No. 3A., 15, Bellamy Papers.
3. Bellamy, "Notebook," Bellamy Papers.
4. See Richard Judd, *Common Lands, Common People: The Origins of Conservation in Northern New England* (Cambridge, 1997), 197–228, for a discussion of the impact of urban tourists and sportsmen on traditional rural society.
5. For an example of the abandoning of farms theme in New England farm journals, see the letter of H. F. French to the *New England Farmer*, Jan. 1853, 13, or William Seward's letter in the same issue, 46–47; see also the *New England Farmer*, Mar. 1854, 126. See also the *Connecticut Valley Farmer and Mechanic*, May, 1853.
6. Lewis Stilwell, *Migration from Vermont* (Montpelier, 1948), 64.
7. *New England Farmer*, 23, Jan. 28, 1871.
8. See Jerold Wikoff, *The Upper Valley: An Illustrated Tour along the Connecticut River before the Twentieth Century* (Chelsea, 1985), and Charles Crane, *Life along the Connecticut River* (Brattleboro, 1939), for a description of the turn-of-the-century dairy industry. See also Richard Wilkie and Jack Tager, eds., *Historical Atlas of Massachusetts* (Amherst, 1991), 32.
9. Wikoff, *Upper Valley*, 145. By the 1920s, Vermont alone had over 421,000 cows.
10. Wilkie and Tager, *Historical Atlas of Massachusetts*, 32. Crane, *Life along the Connecticut*, 76, 77.
11. Crane, *Life along the Connecticut*, 73, 76, 77.
12. Ibid., 87.
13. The completion of rail lines into the rural areas opened up distant hill areas to urban tourism. Railroads in coordination with resort owners advertised the healthful air and waters as well as the scenic beauty of the remote country. Wikoff, *Upper Valley*, 146–151.
14. As early as 1882, New Hampshire's State Board of Health was calling "summer resorts" an important part of the state economy. NHSBH, *First Annual Report* (Concord, 1882), 106. See also NHSBH, *Fourth Annual Report* (Concord, 1885), 4:159, 160. See the board's *Fifth Annual Report* (Concord, 1886), 193, for a discussion of the impact of resorts on local agriculture. For maple syrup see Crane, *Life along the Connecticut*, 46, 58. See Judd, *Common Lands, Common People*, 198–220 for a discussion of the emergence of tourism and the romantic ideal to which it appealed. For an example of the employment of locals and the tension between the urban visitor and the local farmers, see William Dean Howells, *A Traveler from Altruria* (Boston, 1894).

15. Crane, *Life along the Connecticut*, 62.
16. Bellamy, "Notebook," Bellamy Papers.
17. CCFG, *Eighth Biennial Report, 1909–1910* (Hartford, 1910), 8.
18. In the original February 27, 1867, agreement between the various New England states' fish commissions, Vermont and Massachusetts would guarantee that fishways would be built on their major rivers, and would establish a hatchery, and Connecticut would build fishways, abolish gill nets, stake nets, and pound fishing at the mouth of the Connecticut, restrict the times fishers could catch incoming migratory fish, and increase the mesh size of nets. MCIF, *Fifth Annual Report, 1871* (Boston, 1871), 5. See also CCFG, *Sixth Biennial Report, 1905–1906* (Hartford, 1906), 8.
19. By the end of the nineteenth century, fishermen were catching over 90,000 shad on the Connecticut. In a biography of Lyman, H. P. Bowditch claimed that by 1900, "nearly all the shad now taken in our streams have originated in state or national hatchery establishments." H. P. Bowditch, "Biographical Sketch of Theodore Lyman III, 1833–1897" (Washington, DC, 1903), 147, manuscript, Museum of Comparative Zoology, Harvard University.
20. By 1870, Seth Green was hatching over forty million shad fry at Holyoke. See MCIF, *Fifth Annual Report*, 5; *Seventh Annual Report, 1873* (Boston, 1873), 10. See also CCFG, *Sixth Biennial Report*, 8, 9.
21. *Sound Breeze*, October 27, 1896. See also the *Hartford Times*, Mar. 21, 1896, for information about the state hatcheries. CCFG, *Second Biennial Report, 1897–1898* (Hartford, 1898), 8; CCFG, *Third Biennial Report, 1899–1900* (Hartford, 1900), 6; CCFG, *Fifth Biennial Report, 1903–1904* (Hartford, 1904), 4.
22. In 1893, shad fishermen caught fewer than 22,000 shad in the Connecticut. By 1897, over 73,000 were caught in the Connecticut, and in 1898, over 93,450 were brought in. CCFG, *First Biennial Report, 1895–1896* (Hartford, 1896), 37; CCFG, *Third Biennial Report*, 7, *Sixth Biennial Report*, 10, *Second Biennial Report*, 8. New England states were also raising and releasing lake salmon, trout, and bass. CCFG, *Fifth Biennial Report*, 19, *Sixth Biennial Report*, 10.
23. Massachusetts and Connecticut continued to battle over fish catches in the 1880s. I would like to thank Richard Judd for calling my attention to this interstate rivalry.
24. CCFG, *First Biennial Report*, 11.
25. CCFG, *Tenth Biennial Report, 1913–1914* (Hartford, 1914), 7.
26. CCFG, *Ninth Biennial Report, 1911–1912* (Hartford, 1912), 25. The commission's *Ninth Biennial Report* was not the first time the commissioners had asked the state to pass antipollution legislation. In their *Seventh Biennial Report*, the commission "again ... urge[d] such legislation as will correct the abuse ... [of refuse coming from manufacturing plants]." CCFG, *Seventh Biennial Report, 1907–1908* (Hartford, 1908), 36. Again in its next report, the commissioners begged the state to pass antipollution legislation. CCFG, *Eighth Biennial Report*, 29.
27. CCFG, *Eleventh Biennial Report, 1915–1916* (Hartford, 1916), 8.
28. CCFG, *Sixth Biennial Report*, 20, 21.

29. CCFG, *Ninth Biennial Report*, 25.
30. MCIF, *Eighth Annual Report, 1874* (Boston, 1874), 26.
31. Bellamy, "Notebook," 2, Bellamy Papers.
32. Connecticut Valley Water Way Board, "Report on an Investigation of the Connecticut River," (Boston, 1913), 37.
33. Ibid., 36.
34. The Turners Falls Company, for example, rebuilt its dam in 1904, and again in 1912, and again in 1913. Each time, it increased the height of the dam in order to generate more horsepower. The electricity generated by these new higher dams was sold to the Amherst Power Company, the Greenfield Electric Light and Power Company, the Amherst Gas Company, the Easthampton Gas Company, and the Franklin Electric Company. The Holyoke Company sold its electricity to the city of Holyoke. Ibid., 37.
35. See William Leuchtenburg, *Flood Control Politics: The Connecticut River Valley Problem, 1927–1950* (Cambridge, 1953), 2–26, and Lee Webb, "Private Power's Vermont Grip: It Was No Accident," Northeast Center for Social Issue Studies (Brattleboro, 1976), for a discussion of the opposition to the utility holding companies.
36. Typical also was the successful doctor, amateur ornithologist, and avid sports fisherman Leonard Samford. More typical of the older members of the fisheries commission was James Bill, a member of the Connecticut Commission of Fisheries and Game since its founding. Bill was an owner of a book and stationary store in Lyme. As a youth, he "was wont to engage in sports afield and afloat [which] brought him near to nature's heart." In 1849, at a young age, he was elected to the state legislature. He was a member, vice president, and president of the New London Agricultural Society, president of the state agricultural society, and member of the state Board of Agriculture. CCFG, *Third Biennial Report*, 11.
37. F. C. Walcott, "Report to the Connecticut Governor" (Hartford, 1921), 7.
38. See Lyman's critique of Connecticut's commissioner, in which he felt that one of the Connecticut commissioners of fisheries was "half-educated" and the other a self-interested gill netter. Lyman Diaries, Feb. 25, 1867, vol. 22. Initially, the commercial fishers were supported by local fishers who caught shad and salmon for their own domestic consumption. These fishers were joined in the early years of the twentieth century by fishers operating shellfish operations. These coastal fishers had an interest in restricting pollution but were often at odds with the commission over restrictions on catch size. CCFG, *Fifth Biennial Report*, 24.
39. See Arthur McEvoy, *The Fisherman's Problem: Ecology and Law in the California Fisheries 1850–1980* (Cambridge, 1986) for a discussion of this problem.
40. CCFG, *Sixth Biennial Report*, 8.
41. Ibid., 6.
42. CCFG, *First Biennial Report*, 7.
43. CCFG, *Ninth Biennial Report*, 9.

44. CCFG, *Eighth Biennial Report*, 9.

45. CCFG, *Sixth Biennial Report*, 6.

46. Ibid. For a discussion of rural resistance, see Edward D. Ives, *George Magoon and the Down East Game War* (Urbana, 1988).

47. Ibid.; MCIF, *Fourth Annual Report, 1870* (Boston, 1870), 42.

48. CCFG, *First Biennial Report*, 7.

49. CCFG, *Second Biennial Report*, 6.

50. John F. Reiger, *American Sportsmen and the Origins of Conservation* (New York, 1975).

51. In 1872, Spencer Baird of the U.S. Commission on Fisheries in a letter to E. M. Stillwell noted that concern over diminished fish was not an issue "so that the sportsman can capture them with the fly or the man of means be able to procure a coveted delicacy in large quantities, and at moderate expense," the point was to have the numbers of fish grow "substantially in order to provide food and employment to the masses." MCIF, *Eighth Annual Report, 1873* (Boston, 1874), 44.

52. Although the New England states established their fish hatchery program to return shad and salmon to the Connecticut and Merrimack, within a few short years they were also stocking trout, bass, and land-locked salmon in inland lakes and streams. These fish were primarily for the sport fishers. In 1870, Connecticut, for example, stocked 9,000 land-locked salmon in state lakes. CCFG, *Sixth Biennial Report*, 11. In the 1890s, fish and game clubs emerged, mostly among wealthy urbanites. These clubs encouraged the fish commissions to shift their attention to game fish and habitat protection. Reiger, *American Sportsmen*, 35.

53. MCIF, *Eighth Annual Report*, 18. Commonwealth of Massachusetts, "An Act For Encouraging the Cultivation of Useful Fishes," Sen. Doc. 91 (Boston, 1866).

54. CCFG, *Eighth Biennial Report*, 7; see also CCFG, *Tenth Biennial Report*, 6.

55. CCFG, *Second Biennial Report*, 6.

56. Ibid.

57. CCFG, *First Biennial Report*, 76.

58. Ibid., 15.

59. Ibid.

60. Ibid.

61. Publications such as *Forest and Stream*, begun in 1873, and *Field and Stream*, begun in 1874, by the turn of the century were focusing on the link between conservation, manhood, and patriotism. It was an ideology that particularly appealed to elite sportsmen like Roosevelt and the members of his Boone and Crocket Club. See Judd, *Common Land, Common People*, for a discussion of the interest of elite sportsmen in conservation.

62. Walcott, *Report to the Connecticut Governor*, 9.

63. Ibid., 10. Walcott, who was later elected to the Connecticut state senate, also noted in his report to the governor that sport hunting is a uniquely American sport, since "so long as the ordinary shooter does not trespass on posted land he has a fundamental right to take this wild game wherever he

finds it. This country has a different law than that of any other country. It is that wild game belongs not to the land owner, but to the people." Walcott, *Report to the Connecticut Governor*, 10.

64. Ibid., 11.
65. CCFG, *Fifth Biennial Report*, 22.
66. CCFG, *Eleventh Biennial Report*, 9.
67. Walcott, *Report to the Connecticut Governor*, 18.
68. In Connecticut, 122 out of 290 of those convicted for fish and game law violations had Italian surnames. Of those convicted, 2 our of every 3 had foreign surnames. CCFG, *Tenth Biennial Report*, 53–63.
69. Walcott, *Report to the Connecticut Governor*, 42.
70. Ibid., 37.
71. Ibid., 20, 44.
72. This new focus on game habitat protection is reflected in the change in names of what had initially been the Massachusetts Commission on Fisheries. In 1870, the commission changed its name to the Massachusetts Commission on Inland Fisheries. In 1886, the name was changed to the Massachusetts Fish and Game Commission. In 1887, it was changed to the Commission on Inland Fisheries and Game, and in 1900, the name was changed to the Conservation Commission.
73. Walcott, *Report to the Connecticut Governor*, 20, 21.
74. Quoted in CCFG, *First Biennial Report*.
75. CCFG, *Second Biennial Report*, 9. The increase focused on the advantages of sports hunting and fishing for the economy and tourism also reflected a general growth in the interest of tourism as a source of income. Pollution reformers and conservationists began to argue that the region's environment needed to be protected in order to maintain the tourist economy and the infusion of tourist dollars into these rural areas. As the New Hampshire State Board of Health noted in 1882, since tourism and "summer resorts" had become important to the state's economy, "it becomes then a duty for us to foster by every known means all of the legitimate methods that will preserve the present reputation." See NHSBH, *First Annual Report*, 1:106, *Fourth Annual Report*, 4:159–160.
76. CCFG, *Fourth Biennial Report*, 10. In defending game restriction, the commission argued in the same report that the people of the state have realized that "bountiful as nature has been we must not abuse her generosity, or we shall forfeit one of her kindest gifts." When the legislature passed a bill restricting the killing and selling of partridges, quail, and wood duck for two years, the commission felt that it was "as if in answer to the voice of nature appealing to us to spare . . . a few and preserve them before it was too late" (10).
77. Walcott, *Report to the Connecticut Governor*, 18. In earlier reports, the commissions on inland fisheries depended upon old-timers, usually rural farmers, or established commercial fishers as their sources of information. In its first biennial report, the CCFG had a different source of information. "Any observant sportsman knows that at the end of each five year period there

is less game in any given section in this state with which he is familiar." CCFG, *First Biennial Report*, 9. And in 1921, Walcott justified game preservation by reporting to the governor that "the prevailing opinion among the sportsmen of the state is that the fish and game are disappearing." *Report to the Connecticut Governor*, 7.

78. Walcott, *Report to the Connecticut Governor*, 17. See Judd, *Common Lands, Common People*, and Ives, *George Magoon*, for a discussion of the class-based nature of these restrictions.

79. Walcott, *Report to the Connecticut Governor*, 18.

80. CCFG, *Sixth Biennial Report*, 37.

81. CCFG, *Eighth Biennial Report*, 26.

82. CCFG, *Fourth Biennial Report*, 10.

83. Walcott, *Report to the Connecticut Governor*, 10.

84. CCFG, *Twelfth Biennial Report, 1917–1918* (Hartford, 1918), 6.

85. CCFG, *Eleventh Biennial Report*, 6.

86. Walcott, *Report to the Connecticut Governor*, 10.

87. See Ives, *George Magoon*.

88. CCFG, *A Report of Investigation Concerning Shad in Connecticut* (Hartford, 1925), 9.

89. Ibid., 7.

90. Ibid., 21, 22, 26.

91. Ibid., 8–9.

92. See Reiger, *American Sportsmen*, for a discussion of the role of sportsmen and their various organizations in defending the conservation idea.

CHAPTER 11

1. Lyman Scrapbook.

2. *Boston Herald*, July 24, 1882.

3. Lyman's letter to the *Chronicle*, Oct. 7, 1882, Lyman Scrapbook.

4. The most dramatic drop in mortality occurred at the end of the nineteenth century, especially for the under-thirty segment of the population. Massachusetts's mortality rates grew in the 1870s and then slowly declined from the 1890s to the 1930s, with the exception of the influenza spike of 1918–1919. Edger Sydenstricker, *Health and Environment* (New York, 1933), 151–153. Massachusetts experienced dramatic declines in typhoid fever and diphtheria beginning in the 1880s. By 1920, typhoid had been effectively eliminated as a serious killer in the state, while deaths from diphtheria were cut by 95 percent. Reflecting on these statistics, Sydenstricker noted that "hours of labor were long, housing was grossly insanitary, and cities were without adequate means of disposing of human excreta and for providing unpolluted water supplies" (183–184).

5. Ibid., 183. Sydenstricker went on to note optimistically that "so much progress has been made during the last half century that we now seem to stand on the threshold of accomplishments so far-reaching and so great as to defy the imagination" (184).

6. It is important to remember that since the earliest years of settlement, the communities relied on two institutions to protect them and their rights: One was the collective entity of either the town or the general court, and the other was the right to take either direct action or tort action against the individual or entity that had damaged your property rights. In a rural society, the balance between protecting the good of the whole and protecting the good of the individual seemed easy to maintain. In protecting one's individual property rights, one was also protecting a well-ordered society. Increasingly in the nineteenth century, the maintenance of personal rights and property rights became more complex. The agent of the violation of one's rights became more diffuse, and the questions of what one had a right to and who had rights became more difficult to determine. If a polluter dumped sewage in a river, was only the property owner downstream damaged? As these issues arose, more and more New Englanders came to believe that the traditional tort action would not suffice to protect the rights of the community. In that context they claimed not only a right to a clean environment, but the right and indeed obligation of the state to protect that right.

7. In 1835, in *Palmer v. Ferrill*, the court ruled that "contingent, remote and indirect" effects either beneficial or negative could not be charged against a corporation. *The Palmer Company Petitioners v. Isaac Ferrill* (Sept. 1835), MA Reports, Pickering, 17:58. *Royal Call v. The County Commissioners of Middlesex Country* (Oct. 1854), MA Reports, Gray, 2:232.

8. MSBH, *Sixth Annual Report of the State Board of Health, 1874* (Boston, 1875), 354.

9. NHSBH, *First Annual Report of the NHSBH* (Concord, 1882), 1:6.

10. It is ironic that Massachusetts, which was a forerunner in antipollution legislation, should have a representative in Congress, Allen Treadway, who in 1921 opposed a proposed ban on throwing acids into streams because it would hurt the paper interests of the Connecticut River Valley. William Leuchtenburg, *Flood Control Politics: The Connecticut River Valley Problem, 1927–1950* (Cambridge, 1953), 24–25.

11. CSBH, *Third Annual Report of the State Board of Health, 1878* (Hartford, 1880), 139.

12. Deaths from typhoid per 10,000 inhabitants dropped in Holyoke, for example, from 23.3 in the period 1871–1875 to 3.8 for 1891–1895. Chicopee saw a similar decline, as did Springfield. MSBH, *Twenty-eighth Annual Report of the State Board of Health, 1897* (Boston, 1897), xiv.

13. In a 1938 Works Progress Administration report on the Connecticut River Basin, the investigators noted that "as a result of excellent game stocking and wildlife management in the extensive state-owned properties large numbers of wildlife . . . are to be found. Fishing conditions in the watershed are exceptionally good on the ponds and smaller streams due to the extensive stocking program." WPA, "Drainage Basin Study, number 2: Connecticut River" (Boston, 1938), 102.

14. MSBH, *Sixth Annual Report*, 354; NHSBH, *First Annual Report*, 5:13.

15. Edward Everett Hale, *Tarry at Home Travels* (New York, 1906), 95. CSBH, *Twenty-Fourth Annual Report of the Connecticut State Board of Health, 1901* (Hartford, 1902), 261–264.

16. James Olcott, speech before the Agricultural Board of Connecticut, reprinted in CSBH, *Ninth Annual Report of the State Board of Health* (Hartford, 1887), 239, 241, 242.

17. See CSBH, *Twenty-second Annual Report of the State Board of Health, 1899* (Hartford, 1900), xix. Bowditch claimed that the Massachusetts board's reports stimulated other states, particularly urban-industrial ones, to form boards. Bowditch Papers, 8:19. Charles Chapin of Providence, Rhode Island, not only was in communication with the public health activists in Massachusetts, but became an active member of the Massachusetts Association of Boards of Health.

18. Theodore Lyman III Papers, vol. 48, Lyman Papers. H. P. Bowditch, "Biographical Sketch of Theodore Lyman," manuscript, Comparative Museum of Zoology, Harvard; NHCF, *Report Made to the Legislature of New Hampshire, June 1872* (Concord, 1872).

19. New Englanders had been behind the push for a national board of health. Bowditch had written an appeal to Congress and the president asking for such a board. The New Hampshire State Board had also argued that there needed to be a national board. NHSBH *Fifth Annual Report* (Concord, 1886), 5:142.

20. *Bowditch Correspondence*, 2:245.

21. Oscar Handlin and Mary Flug Handlin, *Commonwealth: A Study of the Role of Government in the American Economy: Massachusetts, 1774–1861* (Cambridge, 1961). See also Barbara Rosenkrantz, *Public Health and the State: Changing Views in Massachusetts, 1842–1936* (Cambridge, 1972).

22. WPA, "Drainage Basin Study," 97. The WPA study went on to note that the Springfield Chamber of Commerce described the river as "not desirable for public bathing nor well suited for fish life" (97). As late as 1938, out of a total of forty-six cities and towns discharging into the river, representing almost a third of a million people, only two towns, Amherst and Easthampton, had sewage treatment plants, and Easthampton's was inadequate (97). Indeed, the river was so badly polluted by sanitary and industrial wastes that the WPA found it not only too polluted for swimming, but even boating, canoeing and fishing were "distinctly repugnant" (132). And into the 1930s, Massachusetts still continued to struggle with legislation to stop pollution. In 1940, when the WPA relooked at the Connecticut, it noted little improvement. WPA, "Report, 1940" (Boston, 1940), 256.

23. Federal Security Agency, Public Health Services, "Summary Report on Water Pollution: New England Drainage Basins" (Washington, DC, 1951).

24. Leuchtenburg, *Flood Control.*

Index

Abbott, Jehiel, 73
abolitionism, 104
   Henry Bowditch and, 104, 105–106, 129, 229n24
Act for Encouraging the Cultivation of Useful Fishes (Mass., 1869), 97–98
Act Relative to the Pollution of Rivers, Streams, and Ponds (Mass., 1878), 117
Act to Protect the Purity of Inland Waters (Mass., 1886), 135
Adams, Henry, 81
Adams paper mills, 59–60
Addams, Jane, 159
Agassiz, Alexander, 80, 220n16
Agassiz, Louis, 80, 81, 86, 106, 224n71
Agawam, Mass., 36
Agawam River, 64
Agricultural Board of Connecticut, 49
agricultural reformers, 13, 20–21, 22–23, 80–81, 86
air pollution, 40, 60, 191
alewives, 24
Aluminum Smelting Company, 157
Alvord, Justin, 83
Americanism, 172–173
American Medical Association (AMA), 73, 103, 104
American Public Health Association, 123, 144, 187
American Statistical Association, 73
Ames, David, 37

Ames, John, 37
Ames, Nathan, 36
Ammonoosuc River, 40, 58
anadromous fish. *See* migratory fish
Andrew, Gov. John, 79, 103
anglers. *See* sports fishers
"anticontagionist" theory of disease, 73
antimodernism, 3, 126
Army Corps of Engineers, 8
Ashuelot River, 53
Atwood, N. E., 87

Baird, Spencer F., 95, 187
Baker, Edmond, 26
Banks, Nathaniel, 73
bass, 15, 170, 174, 251n52
bathing, 151–152, 255n22
Beard, Daniel Carter, 172
Bellamy, Edward, 161, 162, 164
Bellows, Judge Henry Adams, 88, 93, 94, 224n67
Bellows Falls, 22, 27, 84, 222n43
   canal at, 28
   factories at, 36, 37, 55, 56
Bellows Falls Company, 20, 34
Bemis, Silas, 66–67, 69
Berlin, Ct., 36
Bigelow, Chief Justice George, 119
Bigelow, Richard, 18, 196n41
Biglow, Isaac, 34, 35
Bill, James, 250n36
biochemical oxygen demand (BOD), 149
Blake, Bill, 36–37

257

boardinghouses, 43–44, 47, 52
Boardman, Elijah, 64
boards of health. *See* state boards of health
Boston, Mass., 5, 36, 50, 51, 73
Boston Associates, 36, 42, 100, 219n6
Boston Courier, 104
Boston Globe, 124
Boston Herald, 124
Boston Medical and Surgical Journal, 123
Bowditch, Dr. Henry Ingersoll, 103–107, 122, 227–228n2
  as abolitionist, 104, 105–106, 129, 229n24
  as advocate of strong role for state, 111, 112–115, 118, 124, 127–128, 160, 187
  as chair of state board of health, 103, 106–107
  on citizens' right to a clean environment, 111, 114, 115, 118, 191
  and eclipse of state board of health, 120–121, 124
  influence on, of English reforms, 113–114, 129
  radicalism of, 6, 104, 106, 113, 188, 228n13
  on role of science, 118, 128
  willingness of, to confront manufacturers, 118, 121, 127–128, 182, 191
Bowditch, Olivia Yardley, 105
Boy Scouts, 172
Brandeis, Louis, 159
brass industry, 59
brickmaking, 45
Brockton, Mass., 138–139
Brookline v. Mackintosh, 123
Burnell, Joseph, 34
Burnside, Ct., 150–151
Butler, Benjamin, 120, 122, 124

Call, Royal, 111, 145, 182
canal dams, 27–30
  flooding caused by, 29–30
  lawsuits over, 29–30
  use of, for waterpower, 30

canals, 27–30
  funding of, 28–29, 30
  see also canal dams
cancer, 244n88
capitalists. *See* investors; manufacturers
carding and picking machines, 34
Carson, Rachel, 171, 191
Case Metal Works, 157
cattle, 14, 19–20
Chadwick, James, 81
Chadwick, Sir Edwin, 61, 113, 129, 217n58
Chapin, Cyrus, 18
Chapin, Enoch, 65
Chapin brothers (William, Levi, and Joseph), 34, 50
Chapman, R. A., 86
charters, ancient, 15, 96–97, 200–201n105
Chase, John, 43
Chase, Stewart, 224n70
Chemical Paper Company, 55
Chickering, Jesse, 38
Chicopee, Mass.
  diseases in, 109, 137–138, 210n85
  industrial pollution from, 144, 150
  rise of, as manufacturing center, 36, 40, 52, 58, 205n17
  sewage disposal in, 52, 53, 109, 142, 143, 144
Chicopee River, 50
  damming of, 36
  pollution of, 142, 143
cholera, 47, 73, 109, 133, 134, 135, 145, 182
Christ, Ernest Wilson, 155
Civil War, 80, 129
Clark Carriage and Wagon shops, 36
Clean Waters Act of 1967, 189
Clifford, Justice Nathan, 99
coal, 40, 207n44
coastal cities, 50, 51. *See also* Boston, Mass.
colonial era, 24–26. *See also* charters, ancient
common law, 64, 65, 214n9
Commons, John R., 159–160

Commonwealth v. Essex Company, 68, 86, 89, 99
Company for Rendering the Connecticut River Navigable by Water-Quechee, 29
Company for Rendering the Connecticut River Navigable by Bellows Falls, 28, 36
Concord River, 117
Connecticut Chemical Company, 156
Connecticut Commission of Fisheries and Game, 100, 155–156, 166–167, 169, 173, 175–177
   and sports fishing, 166–170, 175–176
   and stream pollution, 148, 153, 155–156, 165
   see also Connecticut Board of Industrial Wastes
Connecticut Grange, 153
Connecticut Industrial Wastes Board, 154–158
Connecticut Manufacturers Association, 154
Connecticut River, 6–8, 11–12, 117
   bathing in, 151–152, 189, 255n22
   dams on, 27–30, 35, 42 (*see also* Holyoke dam)
   early abundance of fish in, 15–16, 24, 70, 84, 200n100, 222n43
   falls on, 27, 29 (*see also* Bellows Falls; Hadley Falls; Turners Falls)
   fish restoration in, 164–165, 177, 189
   industrial wastes in, 117, 144, 255n22
   navigation on, 27–29
   sewage in, 61, 138, 142, 143–144, 255n22
   today, 189
   tributaries of, 23 (*see also* specific streams)
   see also migratory fish
Connecticut River Power Company, 166
Connecticut State Board of Health, 59, 111, 131, 140–141, 158–159
   on causes of diseases, 51, 60, 134–135, 138, 139, 239n15
   creation of, 132, 232n85, 233n86
   Hartford investigations by, 51, 138, 239n15
   on hazards to fish, 53, 59, 61, 149
   on industrial wastes, 59, 140–141, 148–149, 153–158
   on need for strong role for state, 111, 114, 116, 143
   New Britain investigations by, 52, 53, 59, 107–108, 140–141
   recommendations of, for sewage disposal, 143, 144
Connecticut State Water Commission, 158
Connecticut Valley Farmer and Mechanic, 21
consumption (tuberculosis), 47, 73
Cook, Stephen, 34
corporations, 28, 183–184
Cotton, Job, 34, 35
cotton textile factories, 34, 37, 43–44, 57
country stores, 16–20
courts, 66–69, 119. *See also* lawsuits; Massachusetts Supreme Judicial Court
Cowell, William, 67
Crosby, A. H., 51
Curtis, Joseph, 73

dairy farming, 22, 162–163, 207–208n49
   and pollution, 110, 163
dams
   for electric power, 166
   for factories, 24, 28, 29–30, 42–43, 127 (*see also* Holyoke dam)
   and fish migration, 25–27, 63, 64–66, 77, 83–84, 88, 167 (*see also* fishways)
   flooding of fields by, 29–30, 64, 67, 68–69, 203n139
   sabotage of, 63–64, 65, 111
   for small, traditional mills, 23, 24, 26–27, 63, 77
   see also canal dams
Davis, Dr. Robert, 107, 129
Deerfield, Mass., 23
Deerfield River, 166, 243n84

Index 259

deforestation, 12, 22, 39–40, 56, 75, 167
  and fish decline, 75, 222n40
  and stream flow, 22, 40, 75
depression of 1816, 35, 36
Derby, Dr. George, 103, 107, 229n30
Dewey, Albert, 37
diarrhea, 109, 133, 239n16
diphtheria, 47, 109, 239n15, 253n4
diseases, 29, 47, 60, 73, 145, 182
  causation of, 60, 72–74, 108–109, 134–136 (*see also* germ theory)
  see also specific diseases
dissolved oxygen (DO), 83–84, 91, 149–151, 152
doctors, 72–74, 125. *See also* Bowditch, Dr. Henry Ingersoll; Walcott, Dr. Henry P.
Donnelly, Charles Francis, 120–121, 124, 125, 145, 190
Douglass, Frederick, 104
drinking water, 47–48, 138, 146
  filtration systems for, 137, 138–140, 241n49
Drown, Thomas, 135, 136
Dwight, Edmund, 36, 42, 205n16
Dwight, Theodore, 22, 35, 61, 63
Dwight, Timothy, 11–14, 15, 31, 45, 61
  changes in Connecticut Valley noted by, 4–5, 11, 13, 22, 23, 31, 35, 63
dysentery, 47, 73, 109, 133, 135, 145, 182

eels, 15
electric power, 165–166, 250n34
Eliot, Charles, 80
Ely, Joseph, 83
Ely, Samuel, 42, 77, 182
Emerson, Elizabeth and Hannah, 129
Emerson, G. B., 40
Emery Wheel Factory, 157
Enfield, Ct., 36
  Connecticut River rapids at, 12, 27
England: public health in, 113, 232n81
  U.S. health reformers' interest in, 113–114, 116, 129, 137, 232n82

Essex case. See *Commonwealth v. Essex Company*
Essex Company, 84–86, 89, 133, 222n46, 224n70

factories, 31, 42–45, 53–60
  sewage from, 46, 50–52, 53
  workforce of, 37
  working conditions in, 47, 145
  see also industrial wastes
farmers, 13–23
  and country stores, 16–20
  and fish, 15, 16, 24, 71, 74–75, 77
  and flooding by dams, 28, 29–30, 64, 68–69, 203n139
  gentlemen, 80–81, 86
  hill-country, 13–14, 21, 39, 162, 163–164, 198n75
  lowland, 13, 23, 38–39, 162
  market pressures on, 17–18, 19–23, 38–39, 162–163
  and pollution, 58, 163
  see also dairy farming
Federal Power Commission, 8
Ferrill, Isaac, 39
Field, Alfred, 93
filtration systems, 137, 138–140, 241n49
fish
  as food, 15, 16, 26, 71, 90, 95–96, 98, 169, 170, 196n32, 200n100, 251n51
  as indicators of a river's health, 7–8, 141
  and industrial wastes, 59, 60, 91–92, 140, 153, 156, 165, 170
  and sewage, 92, 141
  see also migratory fish
fish and game clubs, 251n52
fish and game commissions. *See* state fish and game commissions
fish breeding (pisciculture), 77, 87–88, 93, 164–165, 223n61, 249n19, 251n52
fishers, 15–16
  commercial, 54–55, 170
  farmers as, 54–55, 170
  sports, 71, 95, 167, 169–170
  see also fish: as food

fishways, 92–93
  controversy over, 79, 84–86, 98–99, 225n97
  limited success of, on big dams, 100, 227n129
  and older, smaller dams, 25–27, 77
Flagg, Samuel, 64
flashboards, 68
flax, 20, 21, 194n16
flooding (by dams), 29–30, 64
  lawsuits over, 29–30, 64, 68–69, 203n139
  and Mill Acts (Mass.), 28, 64, 68–69, 183
floods, 22, 189
Folson, Charles, 117, 135
forests, 12–13, 56, 163–164
  cutting of, for lumber, 22, 31, 39–41
  farmers and, 22, 39
  see also deforestation
Francis, James, 90, 94, 133, 224n72
Franconia, N.H., 37, 58, 59
French River, 243n84
Fugitive Slave Law, 104
Fuller, George, 139

game, wild, 15
  decline of, 77, 193n9
  see also hunting
Garrison, William Lloyd, 104, 105
gentlemen farmers, 80–81, 86
germ theory, 134–136, 139, 140, 142, 239n20
Gilmore, Gov. Joseph, 88, 224n67
Grange, the, 153
Gray, Asa, 81
Gray, Justice Horace, 227n126
Green, Copley, 105
Green, Seth, 93, 99
gristmills, 24, 26
Gully Brook, 53

Hadley, Mass., 23, 181
Hadley Falls, 15
  dams at, 28, 29–30, 42–43, 127
  factories at, 37, 42–45
  see also Holyoke, Mass.
Hadley Falls Company, 37, 38, 39, 42–44, 52

Hagen, Hermann, 81
Hale, Edward Everett, 56, 186
Hamden, Ct., 157
Hampden Mill, 44
Hampshire County, Mass., 162
Hampshire Gazette, 21, 33, 38–39
Handlin, Mary and Oscar, 187
Hartford, Ct., 27, 151, 178
  diseases in, 109, 138
  downstream pollution caused by, 53, 108, 150
  drinking water in, 138
  as industrial city, 36, 40, 58
  sewage disposal in, 53, 108, 143–144, 150, 230n44
Hartford, Vt., 37
Hartford Daily Courant, 153
hatcheries, 93, 164–165, 249n19, 251n52
Hatfield, Mass., 23
Hays, Samuel P., 187
health reformers
  belief of, in citizens' right to a clean environment, 3, 111, 114, 115, 118, 191, 238n7
  as pioneer advocates of an activist state, 3, 62, 103, 111, 112–116, 118, 124, 127–128, 131, 143, 160, 185, 187
  see also Bowditch, Dr. Henry Ingersoll; state boards of health; Walcott, Dr. Henry P.
hemp, 21
Henle, Jacob, 108, 134
herring, 15
Hill, Octavia, 113
hill farms, 13–14, 21, 39, 162, 163–164, 198n75
Hoadley, John, 229n31
Hockanum River, 53, 59–60, 107, 150–151, 155, 241n50
Holcomb, Gov. Marcus H., 154, 155
Holmes, Oliver Wendell, 81, 105
Holyoke, Mass., 40–48
  diseases in, 47, 48, 52, 109, 254n12
  drinking water in, 47, 48
  factories in, 42, 43–45, 55, 99
  industrial pollution from, 46, 48, 144, 150

Index 261

Holyoke, Mass. (*continued*)
  rapid growth of, 45–47
  sewage disposal in, 46, 47, 53, 138, 142, 143, 144
  see also Holyoke dam
Holyoke dam, 41, 42–43, 143
  conversion of, for electric power, 166
  and fish migration, 75, 88, 98–99
Holyoke Lumber Company, 41
Holyoke Water Power Company, 41, 44–45, 98–99, 166, 227n126
Hoosick River, 243n84
Hop Brook, 53
Howard, Otis, 68
Hubbard, Justice Samuel, 24
hunting, 75, 172–173, 251–252n63

immigrants, 74, 252n68
individualism, 125, 126
industrialists. *See* manufacturers
industrialization, 34, 126–127
  competing views of, 4–6, 63, 71–72, 126–127, 190–191
industrial wastes, 53–60
  downplaying of, with rise of germ theory, 134, 140
  and fish, 59, 60, 91–92, 140, 153, 156, 165, 170
  harm caused by, to downstream manufacturers, 59–60, 119
  health reformers' efforts to regulate, 131–133, 146, 147–151, 153–154, 159
  manufacturers' resistance to regulation of, 98, 132–134, 153, 154, 156, 158, 165
  pollution of streams by, 53–60, 91–92, 119, 146, 147–151, 243n84
  sources of, 53–60
  and water recreation, 151–152
  see also stream pollution
Industrial Wastes Board (Ct.), 154–158
International Paper Company, 56
investors, 5, 36. *See also* Boston Associates
"invisible hand," 50, 71

Ireland Parish, Mass., 45. *See also* Holyoke, Mass.
ironworks, 36, 37, 58–59. *See also* Stanley Iron Works

Jarvis, Dr. Edward, 73, 103
Jefferson, Thomas, 34
Judd, Sylvester, Jr., 5, 14, 15, 20, 84, 181
Judd, Sylvester, Sr., 19–20

Keene, N.H., 53
Kingsley, Ebenezer, 14
Kirkwood, James, 116, 133
Koch, Robert, 108, 134

Lawrence, Mass.
  dam at, 84–86, 98
  diseases in, 109, 133, 137–138
  factories in, 42, 79
  public-health research in, 136–139
lawsuits
  over flooding caused by dams, 29–30, 64, 67, 68–69, 203n139, 215n26
  over pollution, 110, 111, 119–120, 123, 140, 144–145, 152
  see also Massachusetts Supreme Judicial Court
leather industry, 58
Lebanon, N.H., 37
Lebanon Falls (Olcott's Falls), 27, 29
Lisbon, N.H., 60
Lister, Joseph, 108, 134
Littleton, N.H., 58
livestock, 14, 19–20, 194–195n17
Locke, John, 25
Lombard, Nathan, 119
Loring, John, 24
lotteries, 28, 30, 127
Louis, Dr. P. C. A., 105
Lowell, Mass., 42, 73, 79, 98, 137–138, 211n17
Lowell, Sylvanus, 24, 69
Lowell Gas and Light Company, 129
Ludlow, Mass., 142
lumber, 12, 22, 26, 28, 36, 39
Lydall Brook, 156
Lyman, Elizabeth Russell, 80
Lyman, Theodore (grandfather), 80

Lyman, Theodore, II, 80
Lyman, Theodore, III, 79–83, 187, 220n16
  background of, 5, 41, 79–82, 182, 219n6
  campaigns of, for Congress, 79, 81–82, 145, 181
  faith of, in science and technology, 6, 80–81, 86, 91–92, 99–100, 128, 190
  hopes of, for fish restoration, 93–100, 176
  and industrial wastes, 60, 91–92, 177
  investments of, 5, 79
  mixed praise of, for industrialization, 5, 41–42, 45, 63, 145, 181, 182, 190
  as sportsman, 82, 169
  as state fish commissioner, 60, 79, 83, 90–100, 118

Maine, 70
Manchester, Ct., 60, 150–151
Manchester, N.H., 42
Manhan River, 47
manhood, ideology of, 171–172
manufacturers
  conservationists' reluctance to challenge, 154–158, 178–179
  embracing of germ theory by, 140
  favoring of, by courts, 66–69, 119, 186
  and fish restoration, 98–99, 153, 165, 168
  resistance of, to state regulation, 6, 117–118, 123–124, 126–128, 133, 153–158, 165, 189–190, 247n81
  and science, 128–129
  self-image of, 5–6, 190
  waste disposal practices of, 60 (*see also* industrial wastes)
manufacturing mills. *See* factories
maple products, 163–164
Marsh, George Perkins, 5, 40, 75–78, 171
  on fish decline, 75, 77–78, 86, 222n40
  on need for state action, 76, 77–78

  on outdoor life and manliness, 75, 95, 171
Massachusetts Bureau of Labor Statistics, 47
Massachusetts Commission on Inland Fisheries, 92–93, 99, 165, 252n72
Massachusetts Committee on Artificial Propagation of Fish, 87
Massachusetts Fish and Game Commission, 174–175
Massachusetts Legislative Joint Commission on Obstructions to the Passage of Fish in the Connecticut and Merrimack Rivers, 84, 90
Massachusetts Medical Society, 124
Massachusetts Public Service Commission, 166
Massachusetts River Fishery Commission, 79, 89
Massachusetts State Board of Health, 46, 48, 60, 108, 119, 129, 238n8
  changing status of, 120–122, 125, 132, 145–146, 238n8
  on citizens' right to a clean environment, 3, 115
  creation of, 103–104, 227n1
  on deforestation, 40
  on disease causation, 109, 134, 210–211n12
  on fish decline, 61
  on industrial wastes, 54, 59, 143, 144, 146, 147
  as model for other states, 187
  on need for strong state action, 3, 103, 114, 185
  scientific studies by, 136–139
  on sewage disposal, 49, 52–53, 54, 142, 143, 144, 210n12
Massachusetts Supreme Judicial Court, 24, 52, 63–65, 66–69, 110, 111, 120, 123, 227n126. *See also* Shaw, Chief Justice Lemuel
Meade, Gen. George, 80
measles, 47
meningitis, 109

Merrifield, William, 119–120
Merrifield cases, 119–120, 144–145, 241n49
Merrimack River, 79, 117, 137–138
metal-working industry, 36, 37, 54, 58–59, 153. *See also* Stanley Iron Works
Middlesex Canal Company, 30
Middletown, Ct., 12, 27, 36, 178
migratory fish
 breeding of, 93, 164–165, 249n19, 251n52
 dams and, 74–75, 83–84, 88, 98, 164–165, 200n100 (*see also* fishways)
 early abundance of, 15–16, 24, 84, 210n100
 as important food, 16, 170, 196n32, 222n43
 and interstate tensions, 79, 88–89, 93–94, 99, 224n78, 225n97
 New Hampshire and Vermont and, 79, 88–89, 92–93, 95–96, 99, 224nn67,78, 225n97, 226n123
 and stream pollution, 84, 177–178, 244–245n5
Mill Acts (Mass.), 24, 28, 29, 64, 67, 68–69, 183
Mill Brook (Worcester, Mass.), 120
Millers Falls. *See* Turners Falls
Millers River, 116
Millers River dam, 222n39
Mill River, 34, 37, 38, 144, 208n60, 243n84
Mills, Hiram, 6, 125, 133–134, 136, 137–138, 238n8
mills, manufacturing. *See* factories
mills, traditional, 23, 24, 26–27, 63, 77, 215nn23,26
 importance of, to local communities, 23
Moody, Levi, 14, 16
Moor, Samuel, 23
Moose River, 40
mortality rates, 47, 48, 108, 131, 182
Moseley, Oliver, 64
Muir, John, 10

municipal sewage
 courts and, 120, 140, 142, 211n17, 231n52, 241nn49,50
 dumping of, in streams, 53, 142–143, 147, 211n17
 and fish, 92, 141
 pollution of Connecticut River by, 47, 52, 53, 108, 138, 143–144, 152, 230n44
 and spread of diseases, 135–142
 treatment of, 138–139, 141–142, 152–153
 see also sewers

Nashua, N.H., 42
National Fish and Fisheries Commission, 187
nativism, 74, 173
Naugatuck River, 247n81
Neponset River, 26
New Britain, Ct.
 downstream pollution caused by, 52, 53, 59, 60, 109, 140–141, 152, 231n52
 factories in, 58, 59 (*see also* Stanley Iron Works)
 sewage disposal in, 52, 53, 152, 231n52, 241n50
New England Anti-Slavery Society, 104
New England Commission of River Fisheries, 9, 167
New England Farmer, 162
New England Regional Planning Commission, 8, 189
New Hampshire, 37, 162
 and fish migration, 79, 88–89, 92–93, 95–96, 99, 224nn67,78, 225n97, 226n123
 forest cover in, 39, 40
New Hampshire Game and Fish League, 95
New Hampshire State Board of Health, 108–109, 134, 137, 230n44
 on citizens' right to a clean environment, 3, 238n7
 creation of, 132
 on "economy of nature," 49–50, 62
 on need for state action, 114–116, 185

Newington, Ct., 108
Nichols, William Ripley, 116
Noroton River, 156
Northampton, Mass., 23, 29–30

odors, 52, 60–61, 108, 109–110
Olcott, James, 49, 53, 60, 62, 120, 145
    as former abolitionist, 49, 129
    on industrial waste, 53, 119
    on need for popular mobilization, 49, 119, 120, 186–187, 188
    on need for state action, 62, 111
Olcott's Falls (Lebanon Falls), 27, 29
Olmsted, Frederick Law, 81
outhouses, 47
overfishing, 167, 168
    safeguards against, 25
overhunting, 193n9
oxygen, dissolved (DO), 83–84, 91, 149–151, 152
oyster beds, 148, 153

Palmer, Mass., 36
Palmer Company, 39, 68–69, 206n35
Palmer Company v. Isaac Ferrill, 68–69, 206n35
paper manufacturing, 36–37, 44–45, 54–56, 209n69
    pollution caused by, 54–55, 56
Parker, Chief Justice Isaac, 64, 65
Parkman, Francis, 81
Park River, 53, 59, 60, 107, 141, 239n15
Parsons, Justice Theophilus, 65
Parsons Paper Company, 44
passenger pigeons, 14
Pasteur, Louis, 108, 134
Patten, Matthew, 23
Patterson, J. W., 86–87, 95
Phelps, Earle, 144, 149, 156, 244n4, 247n62
Phillips, S. H., 85
Pinchot, Gifford, 159, 188
Piper's Brook, 53, 61
    pollution of, 52, 53, 59, 61, 107, 108, 109, 152
pisciculture, 77, 87–88, 93, 223n61

Plunkett, Thomas, 104
pollution. *See* air pollution; stream pollution
potash, 22
privies, 47, 50, 51, 52, 109
Progressivism, 6, 100, 159–160, 188–189
Proprietors of Locks and Canals on the Connecticut River, 28, 42
Proprietors of Locks and Canals on the Merrimack River, 68
Putnam, Justice Samuel, 215n23

Quechee River, 37

railroads, 40, 41, 42, 110, 207n39, 248n13
recreation, 151–152, 246n34
Reed, Alfred A., 90
Republican Party, 80
resorts, 164
Rhode Island, 132
Richards, Ellen. *See* Swallow, Ellen
riparian right, 25, 119–120
Rivers and Harbors Act, 8, 189
River Side Park (Hartford, Ct.), 151
Robinson, Gov. George, 124, 125
Robinson, Henry, 93
Rockville, Ct., 150–151
Rocky Hill, Ct., 178
Rogers, William, 81
Roosevelt, Theodore, 159, 172, 188
Ruggles, Joseph, 64
Ruskin, John, 113
Russell, F. W., 93

salmon
    decline and disappearance of, 70, 83–84, 88
    early abundance of, 15, 24, 70, 170, 200n100, 222n43
    efforts to restore, 92, 174
    life cycle of, 200n103
Samford, Leonard, 173, 174, 250n36
sawmills, 24, 26, 224n71
    circular, 36

scarlet fever, 47, 134
science and technology: faith in, 6, 86–87, 128–129, 164
   and avoidance of class conflict, 99–100, 155, 157–158, 159–160, 190, 191
   and fish restoration, 164
   and Progressivism, 159–160, 188
   see also under Lyman, Theodore, III
"scientific business management," 176–177
Sedgwick, Judge Theodore, 24, 64
Sedgwick, William, 135, 136, 138
sewage, 107–108, 135, 147. See also municipal sewage
sewage treatment centers, 138–139, 141–142
   and industrial waste, 152–153
sewers, 108–109, 143, 152, 211n17
   inadequacy of, 47, 52, 108–109
shad, 200n103, 224–225n79
   big dams as obstacles to, 74–75, 98, 164–165, 200n100
   breeding of, 92, 99, 164, 249n19
   early abundance of, 15–16, 24, 84, 210n100
   as important food, 16, 170, 196n32, 222n43
   partial restoration of, 99, 164–165, 177, 249nn19,22
   reduction of, by pollution, 84, 177–178, 244–245n5
   state fish commissions and, 174
Sharon, Mass., 26
Shattuck, Lemuel, 61, 73–74, 123, 217n57, 228n6
Shaw, Chief Justice Lemuel, 24, 66–68, 72, 85–86, 89, 111, 216n26
sheep, 22, 37
Shepard, Levi, 16, 17, 20, 21, 33, 204n3
Shepherd, Col. James, 37–38
silk manufacturing, 37, 54, 58
Silliman, Benjamin, 110
smallpox, 73
Smith, Adam, 50, 62, 71
Smith, Alfred, 62
Smith, Gov. John G., 88
Smith, Hugh, 174

Smith, Jerome Van Crowninshield, 69–71, 72, 75, 87, 178, 190
Smyth, Gov. Frederick, 79, 93
soils, 13, 18–19
South Hadley, Mass., 4, 27–28, 29, 142
South Manchester, Ct., 150–151, 251n50
South Meriden, Ct., 157
sports fishers, 71, 95, 167, 169–170
sportsmen, 75, 172–174, 251–252n61
   and state fish and game commissions, 166–167, 169–170, 171, 173–177, 178
   see also hunters; sports fishers
Spring, Seth, 24, 69
Springfield, Mass., 110
   diseases in, 109, 138
   drinking water in, 47–48
   industry in, 36, 40, 58
   as source of pollution, 138, 142, 144, 150
Springfield, Vt., 58
Stamford Rolling Mills, 156
Stanley Iron Works, 59, 155, 157
   pollution from, 107–108, 155
Stanley Manufacturing Company. See Stanley Iron Works
state: strong role for
   Charles Donnelly's suspicion of, 126
   and fish protection, 76, 131
   George Perkins Marsh's advocacy of, 76–78
   health reformers' advocacy of, 3, 62, 103, 111, 112–116, 118, 124, 127–128, 131, 143, 160, 185, 187
   manufacturers' opposition to, 126–127
   New England's pioneering role in, 131–132, 186–188
state boards of health, 132–133. See also Connecticut State Board of Health; health reformers; Massachusetts State Board of Health; New Hampshire State Board of Health

state fish and game commissions, 164–165, +173+, 249n18
and food fish, 164–165, 167, 169
and sportsmen, 166–167, 169–170, 171, 173–177, 178
and stream pollution, 165, 168, 178–179
see also Connecticut Commission of Fisheries and Game; Massachusetts Commission on Inland Fisheries; Massachusetts Fish and Game Commission
"state medicine," 112–116
steam power, 58
Stevens, Jeduthum, 215n23
stores, country, 16–20
Storrow, Charles, 224n70
Stoughton, Mass., 26
Stowell, Abel, 64
stream pollution
and fish, 84, 91–92, 167, 177–178, 224n71, 244–245n5
rapid increase of, with industrialization, 49–60
see also industrial wastes; sewage
Sumner's Falls, 27, 29
Swallow, Ellen, 135–136
swimming, 151–152, 255n22
Sydenstricker, Edgar, 182

Talbot, Gov. Thomas, 120
tanneries, 16, 36
Tarr, Joel, 159
Taylor, Edmund, 34
technology. See science and technology: faith in
tenement houses, 47, 52
Thayer, Fisher, 67
Thoreau, Henry David, 5, 76, 77, 126
as antimodernist, 3, 71–72, 126, 190–91
and fish decline, 64, 65, 66, 101
Thorndike Manufacturing Company, 36
Thurston, James, 94
Tisdale, Samuel, 97, 176
tool manufacturing, 204n7
tourism, 164, 252n75

"Town Brook Sewer" (Keene, N.H.), 53, 108
Townsend, Charles, 174
Transcendentalism, 72, 217n54
trout, 165, 170, 174, 251n52
tuberculosis, 47, 73
Turners Falls, 27
dam at, 28–29, 83, 200n100, 250n34
typhoid fever, 73, 109, 133, 134, 145, 240n34, 254n12
decline of, 182, 253n4, 254n12
epidemics of, 52, 73, 103, 109, 137–138, 210n85
and sewage-disposal practices, 47, 51, 52, 109, 135, 137–138

Upham, Jacob, 66
U.S. Commission on Fisheries, 95
U.S. Public Health Service, 188
U.S. Supreme Court, 99

Vermont, 166
and fish migration, 74–75, 77, 88, 93, 99
forest cover in, 39, 40
rural changes in, 22, 37, 75, 162, 163
state board of health in, 132
Vermont Republican and American Yeoman, 20–21
Vinton v. Welsh, 15, 202n16
Vose, Daniel, 26

Walcott, Frederic, 166–167, 169–170, 175–176, 178
Walcott, Dr. Henry P., 122–123, 124, 234–235n21, 235n23, 236n40
as head of Massachusetts State Board of Health, 125, 132, 187
manufacturers' hostility to, 124
as president of American Public Health Association, 123, 187, 236n40
Walcott Woollen Manufacturing Company, 66
Walton, Isaac, 70
Warner, Elihu, 15

Index 267

War of 1812, 34
Warren, Mason, 105
waste, human, 50–53, 107–108
   and germ theory, 135, 139
   see also privies; sewage
wastes, industrial. *See* industrial wastes
water closets, 51, 211n16. *See also* privies
water pollution. *See* stream pollution
Weibe, Robert, 187
weirs, 25
Wells, Justice John, 120
Wells River, Vt., 37
Western Railroad, 110
West Springfield, Mass., 36, 142
West Warren, Mass., 52
wheat, 197–198n67
Whipple, George, 139
White, Horace, 64
White River, 40

White River Falls, 55
White River Junction, Vt., 56
Wigglesworth, Edward, 81
William Ashley v. Harlow Pease, 67
Windsor, Vt., 58
Windsor Locks, Ct., 36, 178
Winnepesaukee, Lake, 79
Winnepesaukee River, 79
Winsor, Frederick, 116
wool, 22, 38
woolen factories, 37–38, 41, 54, 56–57
   pollution from, 56–57, 212n45
Worcester, Mass., 120
Works Progress Administration (WPA), 188
Wright, Seth, 17–18

Yardley, Olivia (Olivia Yardley Bowditch), 105
yellow fever, 73